高等院校化学实验室安全风险评估与控制研究成果教材

"十三五"江苏省高等学校重点教材(编号：2020-2-127)

Laboratory Experiments for Inorganic and Analytical Chemistry

无机与分析化学实验
中英双语版

主　编　吴泽颖　丁琳琳　周全法

参　编　（按姓氏笔画排序）
　　　　王文娟　王红松　王晋方
　　　　壮亚峰　向　梅　陈依漪
　　　　吴星星　张震威　魏雪姣

特配电子资源

· 配套课件
· 视频学习
· 拓展阅读

南京大学出版社

图书在版编目(CIP)数据

无机与分析化学实验：中英双语版：汉文、英文／吴泽颖，丁琳琳，周全法主编. —南京：南京大学出版社，2021.7
 ISBN 978-7-305-24796-5

Ⅰ.①无… Ⅱ.①吴… ②丁… ③周… Ⅲ.①无机化学－化学实验－高等学校－教材－汉、英②分析化学－化学实验－高等学校－教材－汉、英 Ⅳ.①O61－33 ②O652.1

中国版本图书馆 CIP 数据核字(2021)第 146364 号

出版发行 南京大学出版社
社　　址 南京市汉口路 22 号　　　邮编 210093
出 版 人 金鑫荣

书　　名 无机与分析化学实验(中英双语版)
主　　编 吴泽颖　丁琳琳　周全法
责任编辑 刘　飞　　　　　　编辑热线 025-83592146
照　　排 南京开卷文化传媒有限公司
印　　刷 南京人文印务有限公司
开　　本 889×1194　1/16　印张 18　字数 550 千
版　　次 2021 年 7 月第 1 版　2021 年 7 月第 1 次印刷
ISBN 978-7-305-24796-5
定　　价 54.00 元

网　　址：http://www.njupco.com
官方微博：http://weibo.com/njupco
微信服务号：njuyuexue
销售咨询热线：(025)83594756

* 版权所有，侵权必究
* 凡购买南大版图书，如有印装质量问题，请与所购
　图书销售部门联系调换

前　言

为了适应高等教育事业的快速发展,满足21世纪高等教育教学改革发展的需要,进一步提高学生动手能力、实践技能及英语水平,以新时代全国高等学校本科教育工作会议精神以及新工科专业建设理念为指导,我们在认真分析国内外同类教材和文献的基础上,着重挖掘降低安全隐患、规避实验风险的关键,对风险进行评估,并找到消除风险的措施,将"安全生产、生命至上"的责任落在实处,同时采用中英文对照的方式,编写了这本《无机与分析化学实验》(中英双语版)教材。

本书内容上以"无机与分析化学实验"基本原理为主线,注重基本操作和基本技能的训练。实验内容分为化学反应原理与物理量测定、元素化学实验和物质的提纯与分析。主要涉及无机盐及配合物的制备、根据化学原理测定平衡常数和以化学分析法对物质进行含量测定。在此基础上我们还编写了综合性、设计性、研究性实验部分,旨在培养学生分析问题和解决问题的能力,锻炼学生独立思考和综合运用的科研能力,增强学生对化学实验的兴趣和探究其内在原理的动力。

本书秉承了现有书籍的精髓和根本,着眼于实验的特点,增加了数字资源(扫码后关注"南大悦学"公众号,即可阅览)和风险评估表。面对大一、大二年级的学生,使他们在预习实验时很直观地知道怎么做实验,面临的风险有哪些,如何避免造成安全事故,以及如果不慎出现风险该如何消除,即使在中学阶段没有做过化学实验的学生,也能清楚地知道用什么仪器、如何操作、如何处理实验数据。

参加本书编写的有丁琳琳(负责第1章、实验1~2、实验13~16及附录的编写),张震威(负责实验21~25的编写),王文娟(负责实验3、6~8、11的编写),向梅(负责实验17~20的编写),陈依漪(负责实验5、9、10的编写),魏雪姣(负责实验4、12的编写),吴泽颖(合编风险评估),王晋方(合编风险评估),王红松(合编风险评估),吴星星(合编风险评估),全书由周全法、丁琳琳和吴泽颖负责统稿。

本书是在化工系和应用化学系各位老师长期努力下完成的。在这里我们感谢学校及相关部门的支持,特别向为我们提供参考书和文献的兄弟院校和老师们表示诚挚的感谢。

由于我们水平有限,书中错误和不足之处在所难免,请读者谅解及批评指正。

编　者
2021年7月

目 录

第 1 章　化学实验基础知识

1.1　基本安全规则 ··· 1

1.1　Elementary safety rules ·· 2

1.2　化学实验室安全知识 ··· 3

1.2　Laboratory safety ·· 5

1.3　实验数据的记录和处理 ·· 7

1.3　Recording and processing of experimental data ·· 8

第 2 章　实验基本操作

实验 1　分析天平的使用及称量 ··· 9

Experiment 1　Operation of analytical balance and weighing ··· 13

实验 2　滴定分析的基本知识和操作 ·· 17

Experiment 2　Elementary knowledge and basic operation of titration analysis ················· 23

第 3 章　化学反应原理与物理量测定

实验 3　摩尔气体常数的测定 ··· 33

Experiment 3　Determination of molar gas constant ·· 37

实验 4　化学反应速率和活化能 ··· 41

Experiment 4　Determination of chemical reaction rate and activation energy ·················· 46

实验 5　弱酸电离常数测定（pH 法） ··· 52

Experiment 5　Determination of ionization constant of weak acid （pH method） ············· 55

实验 6　酸碱反应与缓冲溶液 ………………………………………………………………… 58

Experiment 6　Acid-base reaction and buffer solution ………………………………………… 63

第 4 章　元素化学实验

实验 7　硼碳硅氮磷 ………………………………………………………………………………… 68

Experiment 7　Boron, carbon, silicon, nitrogen and phosphorus ……………………………… 76

实验 8　氧硫氯溴碘 ………………………………………………………………………………… 86

Experiment 8　Oxygen, sulfur, chlorine, bromine and iodine ………………………………… 91

实验 9　钛、钒、铬和锰的化合物 ……………………………………………………………… 97

Experiment 9　Compounds of Ti, V, Cr and Mn ……………………………………………… 106

实验 10　铁、钴、镍的化合物 ………………………………………………………………… 116

Experiment 10　Compounds of Fe, Co and Ni ………………………………………………… 125

第 5 章　物质的提纯与分析

实验 11　氯化钠的提纯 ………………………………………………………………………… 135

Experiment 11　Purification of sodium chloride ……………………………………………… 139

实验 12　硫酸铜的提纯 ………………………………………………………………………… 143

Experiment 12　Purification of copper (Ⅱ) sulfate …………………………………………… 148

实验 13　NaOH 标准溶液的配制与标定及铵盐中氮含量的测定 …………………………… 154

Experiment 13　Preparation and standardization of NaOH standard solution and
　　　　　　　　determination of nitrogen content of ammonium salt ……………………… 159

实验 14　HCl 标准溶液的配制与标定及混合碱的分析（双指示剂法）……………………… 165

Experiment 14　Preparation and standardization of HCl standard solution and
　　　　　　　　determination of the composition of mixed base (double-indicator method) …… 169

实验 15　EDTA 标准溶液的配制与标定及水硬度的测定 …………………………………… 174

Experiment 15　Preparation and standardization of EDTA standard solution and
　　　　　　　　determination of water hardness ………………………………………………… 179

实验 16　EDTA 标准溶液的配制与标定及铅、铋混合液中铅、铋含量的连续测定 ………… 186

Experiment 16　Preparation and standardization of EDTA standard solution and determination of Bi^{3+} and Pb^{2+} in the mixed solution of bismuth and lead …… 191

实验 17　高锰酸钾标准溶液的配制与浓度标定及过氧化氢的含量测定 ……………… 197

Experiment 17　Preparation and calibration of potassium permanganate standard solution and determination of hydrogen peroxide content ……… 202

实验 18　铁矿石中铁含量的测定(无汞法) ……………………………………………… 208

Experiment 18　Determination of iron content in iron ores (mercury-free method) …… 213

实验 19　$Na_2S_2O_3$标准溶液的配制与浓度标定及间接碘量法测定铜盐中的铜含量 …… 219

Experiment 19　Preparation and calibration of $Na_2S_2O_3$ standard solution and determination of copper in copper content salt by indirect iodimetry method ……… 224

实验 20　可溶性氯化物中氯含量的测定(莫尔法) ……………………………………… 231

Experiment 20　Determination of chlorine content in soluble chlorides (Mohr method) …… 235

第 6 章　综合性、设计性、研究性实验

实验 21　硫酸亚铁铵的制备 ……………………………………………………………… 239

Experiment 21　Preparation of ammonium iron(Ⅱ) sulfate ……………………………… 242

实验 22　三草酸合铁酸钾的制备 ………………………………………………………… 245

Experiment 22　Preparation of tripotassium trioxalatoferrate …………………………… 248

实验 23　铬(Ⅲ)配合物的制备和分裂能的测定 ………………………………………… 252

Experiment 23　Preparation of chromium (Ⅲ) complex and determination of splitting energy ……………………………………………………………………… 256

实验 24　氯化一氯·五氨合钴(Ⅲ)水合反应活化能的测定 …………………………… 260

Experiment 24　Preparation of monochloro pentammine cobalt (Ⅲ) hydration and determination of its activation energy ……………………………… 262

实验 25　多金属氧酸盐的制备及其光催化降解性能测试研究 ………………………… 266

Experiment 25　Preparation of polyoxometalates and assessment of its photocatalytic degradation performance ……………………………………………… 268

附 录

附录1　元素的相对原子质量(A_r)表(2011) ··· 270

Appendix 1　Table of standard atomic weights (2011) ······································· 270

附录2　常用酸、碱溶液的物理性质 ··· 272

Appendix 2　Physical properties of common acids and bases ······························· 272

附录3　常用弱电解质的解离常数 ·· 273

Appendix 3　Dissociation constants of common weak electrolytes ························· 273

附录4　常用酸碱指示剂 ·· 275

Appendix 4　Common acid-base indicators ··· 276

附录5　常用缓冲溶液 ··· 277

Appendix 5　Common buffer solutions ·· 278

参考文献 ·· 279

第 1 章
化学实验基础知识

1.1 基本安全规则

所有实验人员必须遵守安全规则以降低实验室事故风险。

1. 仔细计划所有的实验活动。预习实验内容,确保初次实验前,彻底了解实验原理和步骤。
2. 通过阅读风险评估及控制工作表,检查即将在实验中使用的所有化学品的特性和危险。
3. 减少接触危险化学品。避免所有化学品接触眼睛和皮肤,并特别注意呼吸刺激物和吸入危害。
4. 使用任何化学品前应仔细阅读其标签。
5. 任何时候在实验室使用化学品、加热设备或玻璃器皿时都必须佩戴护目镜。
6. 熟悉安全设备的位置,以便在紧急情况下可以快速找到和使用这些设备,并认真学习如何使用实验室的所有安全设备(如洗眼器、安全淋浴器、灭火器等)。
7. 进入实验室需始终穿戴合适的个人防护装备,包括不露脚趾的鞋、长裤、长袖和实验服。
8. 养成良好的化学卫生习惯。不要在实验室里吃东西或用实验室的玻璃器皿喝水。任何食品都不应存放在实验室冰箱中。离开实验室前,务必彻底洗手。
9. 离开实验室时,务必确保锁好实验室门窗。
10. 了解化学品泄漏、火灾和伤害的应急处理方法。
11. 不要用湿手操作电气设备,设备使用中不可无人看管,并需对其进行定期检查。

- 安全规则
- 常用仪器
- 实验操作

1.1 Elementary safety rules

All personnel must follow general safety rules to reduce the risk of accidents in the laboratory.

1. Carefully plan all lab activities. Practice experiments and demonstrations beforehand; make sure you thoroughly understand the science and procedure before performing a lab activity for the first time.

2. Review the properties and hazards of all chemicals you will be using in the lab by reading the Risk Assessment and Control Worksheet.

3. Reduce exposure to hazardous chemicals. Avoid contact of all chemicals with eyes and skin, and pay special attention to respiratory irritants and inhalation hazards.

4. Read all chemical labels prior to use.

5. Goggles must be worn at any time chemicals, heat, or glassware are used in the laboratory.

6. Be familiar with the location of safety equipment, so as to quickly find and use it in an emergency. Seriously learn how to use all safety equipment in the laboratory (e.g., eyewash, safety shower, fire extinguisher, etc.).

7. Wear appropriate personal protective equipment, including closed-toed shoes, long pants, shirts with sleeves and lab coats, at all times.

8. Develop good chemical hygiene habits. Never eat in the lab or drink out of laboratory glassware. No food should be stored or even cooled in the laboratory refrigerator. Always wash your hands thoroughly before leaving the lab area.

9. When leaving the lab, make sure the prep area and laboratory doors and windows are locked.

10. Know emergency procedures for a chemical spill, fire, and injury.

11. Do not operate the electrical equipment with wet hands, do not leave the equipment unattended during use, and check it regularly.

1.2 化学实验室安全知识

1.2.1 个人防护

1. 眼睛保护

① 在实验室中，需要随时佩戴防溅和防冲击的护目镜。
② 彻底洗手之前不要揉眼睛。

2. 服装、头发和首饰

① 应穿能完全遮住脚的鞋子，以免脚部受到地板上的碎屑或碎玻璃和泄漏化学品的伤害。
② 应将长发盘起，避免佩戴领带和围巾，以防止接触化学品、火源及设备。
③ 建议学生摘下首饰，以防止化学品落在首饰上、触电、卡住设备等。

3. 手套

① 如果需要使用手套，请在每次使用前检查手套上是否有孔和裂口。离开实验区域前摘下手套，防止其他表面受到污染。
② 正确清洁或处理手套。

4. 食品和饮料

① 实验室内不允许有食品和饮料。
② 不要在实验室内吸烟或嚼口香糖。

1.2.2 应急程序

1. 疏散

① 当警报响起时，熄灭房间里的任何火焰。
② 迅速离开大楼，关闭身后所有的门窗。

2. 人身伤害

① 简单的急救箱位于各实验室置物柜内。
② 如有必要，将伤者送往急诊室。
③ 化学烧伤：淋浴或使用洗眼水冲洗至少20分钟。必要时脱掉衣服。
④ 出血：清洁伤口，用绷带压迫止血。实施止血者需佩戴乳胶手套。
⑤ 轻微热烧伤（无水泡）：用冷水冲洗烧伤部位至少20分钟。
⑥ 严重热烧伤：不要用水冲洗烧伤部位。用干净、干燥的绷带松松地包扎，拨打120，立即用救护车送往急诊室。
⑦ 中毒：拨打120，立即用救护车送往急诊室。

3. 火灾

① 对于普通可燃材料引发的火灾，如木材、布料、纸张、橡胶和许多塑料，几乎所有类型灭火器均可

以使用,水也是很好的灭火剂。

② 对于易燃液体、气体、油、油漆和油脂引发的火灾,泡沫灭火器、干粉灭火器或二氧化碳灭火器是此类火灾中最有效的灭火器。**不要使用水进行灭火。**

③ 对于通电的电器设备引发的火灾,灭火剂的非导电性很重要,使用二氧化碳或干粉灭火器。**不要使用水进行灭火。**

④ 对于可燃金属,如镁、钛、锆、钠、锂、锌和钾等,引发的火灾,应使用金属灭火剂或沙子。

如果火势太大,无法用灭火器处理,拉响火警报警器并进行疏散。

1.2.3　实验室垃圾

1. 空试剂瓶
将空试剂瓶交由实验室管理人员处置。

2. 锐器(例如,碎玻璃、针头、刀片)
切勿在垃圾桶中放置锋利器物。应将所有锋利器物放到一个防穿刺容器中,并清楚地将标有"锐利"字样的标签贴在该容器外部。如有必要,请在处置前清洁锋利器物。当容器装满时,用胶带密封,交由实验室管理人员处置。

3. 沾有化学品的废纸
如果使用纸巾擦拭泄露的化学品,请将使用过的纸巾放入适当的容器或袋子中,并与常规化学废物一起处置。不要把用于清理化学品的纸巾直接放在垃圾桶里。

• 实验室安全课程

1.2 Laboratory safety

1.2.1 Personal Protection

1. Eye Protection

① Splash and impact protective goggles are required at all times in the laboratory.

② Do not rub your eyes until you have thoroughly washed your hands.

2. Clothing, Hair, and Jewelry

① Wear shoes that completely cover your feet to protect your feet from debris or broken glass on the floor and from chemical spills.

② Constrain long hair, neckties, and scarves to prevent contact with chemicals, fire, and equipment.

③ Students are advised to remove jewelry to prevent trapping chemicals under the jewelry, contact with electrical sources, and catching on equipment.

3. Gloves

① If gloves are required, check for holes and tears before each use. Remove gloves before leaving your work area to prevent contamination of other surfaces.

② Clean or dispose of gloves properly.

4. Food and Beverages

① Food and beverages are not permitted in the laboratory.

② Do not use tobacco products or chew gum in the laboratory.

1.2.2 Emergency Procedures

1. Evacuation

① When the alarm sounds, extinguish any flame that is in the room.

② Leave the building quickly, closing all windows and doors behind you.

2. Personal Injury

① Simple first aid kits are located in the storage cabinets of each laboratory.

② If necessary, transport the injured person to the emergency room.

③ Chemical burns: Use shower or eyewash for at least 20 minutes. Remove clothing if necessary.

④ Bleeding: Clean the wound and apply pressure with a bandage to stop bleeding. If you are helping someone else who is bleeding, use Latex gloves.

⑤ Minor thermal burns (no blisters): Run cool water over burned area for at least 20 minutes.

⑥ Severe thermal burns: Do not run water over burned area. Wrap loosely with clean, dry bandage, call 120 for immediate ambulance transport to the emergency room.

⑦ Poisoning: Call 120 for immediate ambulance transport to the emergency room.

3. Fire

① Fires in ordinary combustible materials, such as wood, cloth, paper, rubber and many plastics. Almost any fire extinguisher is effective, but water is the best extinguishing agent.

② Fires in flammable liquids, gases, oil, paint and greases. Foam, dry chemical or CO_2 extinguishers are the most effective. **Do not use water.**

③ Fires which involve energized electrical equipment where the electrical non-conductivity of the extinguishing agent is of importance. Use Carbon Dioxide or Dry Chemical extinguishers. **Do not use water.**

④ Fires in combustible metals, such as magnesium, titanium, zirconium, sodium, lithium, zinc and potassium. Use metal fire extinguishing agent at safety stations or sand.

If the fire is too large to handle with extinguishers, pull a building alarm activator and evacuate the building.

1.2.3　Laboratory Trash

1. Empty Chemical Bottles

The empty chemical bottles were handed over to the laboratory administrator for disposal.

2. Sharps (e.g., broken glass, needles, blades)

Never put sharps in your trash can. Discard all sharps into a puncture-resistant container that is clearly labels "sharps". If necessary, clean the sharps before disposal. When the container is full, seal it with tape and ask the laboratory administrator to dispose of it.

3. Saturated Paper Waste

If you use paper towels to clean a chemical spill, put them in an appropriate container or bag and dispose of them with your regular chemical waste. Do not put paper towels used for chemical cleanups in your trash can.

1.3 实验数据的记录和处理

1.3.1 实验数据的记录

实验原始数据的记录应是多年后仍可被查阅的永久记录,也是科研和撰写论文的原始资料。学生应在实验过程中养成严谨的科学态度和实事求是的精神,应按照以下要求进行实验数据的记录。

(1) 应使用编有页码的笔记本记录原始数据,不得撕去任何一页。不得将数据记在其他地方事后转抄。

(2) 记录的内容应包括实验过程、现象、仪器、试剂试药及用量等,不得随意拼凑和伪造数据。

(3) 记录实验数据时,保留几位有效数字应和所用仪器的准确程度相适应。为避免修约造成的误差积累,中间步骤应多保留一位有效数字。

(4) 实验中的每一个数据都应记录完整。如需修正,可在不正确的数据上画一条线,然后在上方或旁边写上正确的数据,不可涂改,同时应保持所有数据的可读性。

1.3.2 分析数据的处理

单次测得的一组数据可用 Q 检验法决定可疑数据的取舍,然后计算其算数平均值。用相对平均偏差、标准偏差或相对标准偏差来表示分析结果的精密度。

●视频演示

1.3 Recording and processing of experimental data

1.3.1 Recording of experimental data

The record of the original experimental data should be a permanent record that can still be consulted many years later, and it is also the original data of scientific research and writing papers. Students should develop a rigorous scientific attitude and the spirit of seeking truth from facts during the experiment, and record the experimental data according to the following requirements.

(1) The original data shall be recorded in the notebook with page number, and any page shall not be torn off. Data shall not be recorded in other places and copied afterwards.

(2) The contents recorded shall include the experimental process, phenomenon, instrument, reagent test and dosage, etc. The data shall not be randomly pieced up or forged.

(3) When recording the experimental data, several significant figures should be kept to match the accuracy of the instrument used. In order to avoid error accumulation caused by rounding, one more significant digit should be reserved in the intermediate step.

(4) Every data in the experiment should be recorded. If correction is needed, a line can be drawn on the incorrect data, and then the correct data can be written on the top or beside. It is not allowed to be altered, and the readability of all data should be maintained.

1.3.2 Processing of experimental data

For a group of single measured data, the Q-test method can be used to determine the choice of suspicious data, and then calculate the arithmetic mean. Relative mean deviation, standard deviation and relative standard deviation are used to express the precision of analysis results.

第 2 章

实验基本操作

实验 1 分析天平的使用及称量

【实验目的】

1. 了解电子天平的原理,掌握电子天平的使用。
2. 学会用直接法和减量法称量试样。

【实验原理】

分析天平是化学实验中最重要、最常用的仪器之一。目前常用的分析天平是电子天平。电子天平利用电子装置完成电磁力补偿的调节,使物体在重力场中实现力的平衡;或通过电磁力矩的调节,使物体在重力场中实现力矩的平衡。

【器材与试剂】

1. 器材:电子天平、烧杯、称量瓶、称量纸、表面皿、锥形瓶、药匙。
2. 试剂:粗食盐。

● 视频演示

【实验内容】

1. 电子天平的使用方法

(1) 清洁天平及天平盘:使用天平前需用软刷将天平及天平盘清理干净。
(2) 调水平:查看天平上的水平仪,如果天平不水平,可调节电子天平水平调节脚,使水平仪内空气气泡位于圆环中央。
(3) 预热:电子天平在初次接通电源或长时间断电后再开机时,均应进行预热。每台天平的预热时间各不相同,天平的准确度等级越高,所需预热的时间越长。一般开机预热 20 min 后方可进行称量。

(4) 校准：天平闲置时间较长、位置移动或环境变化后，一般都应对天平进行校准。具体校准操作如下：按校准"CAL"键，天平将显示所需标准砝码重量，放上标准砝码直至出现"g"，校准结束。

(5) 去皮重：置容器于天平盘上，天平显示容器质量，按"去皮"键，显示"0.000 0"时，去皮重结束。

(6) 称量：打开天平门，置被称物于容器中，关闭天平门，天平显示被称物重量，记录结果。

(7) 关机并清洁天平：实验结束清理天平盘，并关闭天平门。

2. 称量练习

(1) 直接称量：对一些不易吸水、在空气中稳定、无腐蚀性的物品，可通过称量瓶或称量纸等称量容器将其置于天平盘上按天平使用方法直接称出其质量。称取称量瓶质量、瓶身、瓶盖的质量，并将数据记录于表1-1。

(2) 减量法称重：使用此方法时应选用开口较大的容器接收试样，以确保试样不会撒落到接收容器外。分别称取 0.4~0.5 g 粗食盐一份，0.2~0.25 g 粗食盐一份，0.10~0.13 g 粗食盐一份，并将数据记录于表1-1。具体步骤如下：

图1-1　减量法

① 将适量试样置于称量瓶中于分析天平上精确称出其总质量 W_1。

② 取出称量瓶，在接收容器上方打开瓶盖，用称量瓶盖轻敲称量瓶的上部，使试样缓慢落入接收容器中，如图1-1，盖妥瓶盖后，方可离开接收容器上方，再将称量瓶及剩余试样放回到天平盘上精确称出其质量 W_2。如此重复，直至倾出的试样质量达到要求。

③ 倾入接收容器的试样质量 W 按以下公式进行计算：$W = W_1 - W_2$。

3. 数据处理

数据处理见表1-1。

表1-1　电子天平称量数据处理

项　目	数　据	
	粗　称	精　称
$m_{称量瓶}$/g		
$m_{瓶身}$/g		
$m_{瓶盖}$/g		
W_1/g		
W_2/g		
W/g		

【注意事项】

1. 天平应安装在稳固、不易震动的实验台上。天平室内应保持清洁干燥，避免阳光直接照射。

2. 应从左右侧门取放称量物，称量物不能超过天平负载。称量易挥发、具有腐蚀性或吸湿性物体时，要盛放在密闭容器中，以免腐蚀和损坏电子天平。

3. 称量物的温度应与室温相同，不得将热的、冷的物品直接放在天平上称量，应待称量物放置室温后再行称量。

4. 天平内应放置干燥剂（常用变色硅胶），干燥剂应定期更换。

【思考题】

1. 分析天平的灵敏度越高,是否称量的准确度就越高?
2. 递减称量法称量过程中能否用小勺取样?为什么?
3. 使用称量瓶时如何操作才能保证试样不致损失?

无机与分析化学实验

风险评估及控制工作表

实验任务	分析天平的使用及称量				评估人	吴泽颖、王红松、吴星星				评估日期	2021年7月1日			
过程步骤	危害品的使用及称量				健康风险		安全措施			清单/风险评估	风险评估			紧急情况/泄露特殊备注
	化合物名称	危害鉴别		化学品安全技术说明书	可能的伤害种类	降低或消除风险	技术	防护眼镜	手套	通风橱	可能性	危害	风险	危害级别
		类别	浓度	用量										高中低
		或物理危害									1～4	1～4		
A: 电子天平的使用方法														
Ⅰ) 清洁天平及天平盘														
Ⅱ) 调水平														
Ⅲ) 预热	电子天平				电击	确保电线已贴上标签	√	√			1	3	3	低	立刻向指导老师或准备实验人员报告
Ⅳ) 校准														
Ⅴ) 去皮重														
Ⅵ) 称量														
Ⅶ) 关机并清洁天平														
B: 称量练习														
Ⅰ) 直接称量：称取称量瓶质量，瓶身、瓶盖的质量，并将数据记录于表1-1。	电子天平	—	—	—	电击	确保电线已贴上标签	√				1	3	3	低	立刻向指导老师或实验准备人员报告
Ⅱ) 减量法称重：分别称取0.4～0.5 g 粗食盐一份，0.2～0.25 g 粗食盐一份，0.10～0.13 g 粗食盐一份，并将数据记录于表1-1。	粗食盐	—	—	1 g	皮肤接触或吞咽有毒	关注	√		√		1	3	3	低	大量水冲洗立刻向指导老师或实验准备人员报告
C: 数据处理														
实验产物														
实验废弃物														
整个过程的最终评估	低风险（按照正确流程操作）													
签名	实验老师：				实验室主任/职业安全健康（OHS）代表：					学生/操作人：				

Experiment 1　Operation of analytical balance and weighing

【Objectives】

1. To understand the principle of electronic balance and master the use of electronic balance.
2. To learn to weigh the sample with direct method and decrement method.

【Principles】

Analytical balance is one of the most important and commonly used apparatus in chemical experiment. The commonly used is electronic balance. It uses electronic device to complete the adjustment of electromagnetic force compensation, so that the object can achieve the balance of force in the gravity field; or through the adjustment of electromagnetic torque, so that the object can achieve the balance of torque in the gravity field.

【Apparatus and Chemicals】

1. Apparatus

Electronic balance; Beaker; Weighing bottle; Weighing paper; Watch glass; Erlenmeyer flask; Medicine spoon.

2. Chemicals

Crude salt.

【Experimental Procedures】

1. Usage of electronic balance

(1) Clean the balance and balance pan: Before using the balance, clean the balance pan and base plate with a soft brush.

(2) Level the balance: Check the level on the balance. If it is not level, adjust the level adjusting feet of the electronic balance to make the air bubble in the level in the center of the ring.

(3) Preheating: The electronic balance shall be preheated when it is powered on for the first time or powered off for a long time. The preheating time of each balance is different. The higher the accuracy level of the balance is, the longer the preheating time is required. Generally, the weighing can be carried out after the machine is started and preheated for 20 min.

(4) Calibration: If lying idle for a long time, position moves or environment changes, calibration

shall be carried out generally. Press the calibration key CAL, the balance will display the required standard weight. Put on the standard weight until the g appears. The calibration is finished.

(5) Tare the balance: Place empty vessel on the pan. Be sure all the enclosure doors are shut. Press the "T" bar. Wait for the display to read "0.000 0".

(6) Measure the sample: Place sample on the pan. Close all the doors of the balance. Record the measurement.

(7) Turn power off and clean the chamber and pan: When weighing is finished, clean out the spills or residue in the pan and the balance. Close all doors of the balance.

2. Weighing practice

(1) Direct weighing: Some stable, water-insensitive and noncorrosive reagents can be weighed out on balance directly. Weigh the mass of weighing bottle, bottle body and cap, and record the data in Table 1-1.

Figure 1-1 Weighing by difference

(2) Weighing by difference: This method assumes that the receiver flask has a large enough opening that there is minimal risk for spillage. Weigh 0.4~0.5 g NaCl, 0.2~0.25 g NaCl and 0.10~0.13 g NaCl respectively. Record the data in Table 1-1. The specific steps are as follows:

① Get a weighing bottle which contains the material you want to weigh out. Weigh the weighing bottle on the analytical balance and record the weight, W_1.

② Transfer material from the weighing bottle into a flask or other receiving container. When transferring, use one hand to hold the weighing bottle slightly tilted while tapping the weighing bottle to allow a small amount of the material to come out. Put the weighing bottle back on the balance pan and record the weight W_2. Repeat the operation until the quality of the poured sample meets the requirements.

③ To get the weight, W, of your transferred material, use the following formula: $W = W_1 - W_2$.

3. Data processing

See Table 1-1.

Table 1-1 Weighing data processing of electronic balance

Weighing project	Data		
	Rough weighing	Precision weighing	
$m_{\text{Weighing bottle}}/g$			
$m_{\text{Bottle body}}/g$			
$m_{\text{Bottle cap}}/g$			
W_1/g			
W_2/g			
W/g			

【Notes】

1. The balance room should be kept clean and dry, and direct sunlight should be avoided. The

balance should be installed on a stable and nonvibrating experimental platform.

2. Get and place the objects from the side doors. The mass should not outweigh the maximum weigh of the balance. Objects with corrosivity and hygroscopicity must be placed in an airtight container before weighing.

3. The temperature of the weighing substance should be the same as that of the room temperature. Hot and cold objects should not be directly put on the balance for weighing. The weighing should be carried out after the weighing object is placed at the room temperature.

4. Desiccant (commonly used discolored silica gel) should be placed in the balance, and the desiccant should be replaced regularly.

【Questions】

1. Is the higher the sensitivity of balance, the higher the accuracy of weighing?
2. Can small scoops be used for sampling in the weighing process of the declining weighing method? Give the reason.
3. How to ensure that the sample is not lost if using the weighing bottle?

Risk Assessment and Control Worksheet												
TASK	Operation and weighing of analytical balance					Assessor	Zeying WU, Hongsong WANG, Xingxing WU		Date	1 JUL. 2021		
Process Steps	Hazard Identification					Health Risk	Safety Measures				Checklist＼Risk Assessment	Risk Assessment			Action
	Chemical Name	Class	Concn.	Amount	MSDS or Physical Hazard	Type of Injury Possible	To reduce or eliminate the risk	Technique	Glasses	Gloves	Fume Cbd.		Likely 1~4	Hazard 1~4	Risk Score	Risk Level H M L	In case of Emergency/Spill Special Comments

Process Steps	Chemical Name	Class	Concn.	Amount	MSDS	Type of Injury Possible	To reduce or eliminate the risk	Technique	Glasses	Gloves	Fume Cbd.	Likely	Hazard	Risk Score	Risk Level	Action
A: Usage of electronic balance																	
Ⅰ) Clean the balance and balance pan																
Ⅱ) Level the balance																
Ⅲ) Preheating	Electronic balance					Electric shock	Ensure cord is tagged	√	√			1	3	3	L	Report immediately to Demonstrator or Prep Room
Ⅳ) Calibration																
Ⅴ) Set the required wavelength on spectrophotometer.																
Ⅵ) Tare the balance																
Ⅶ) Measure the sample																
B: Weighing practice																	
Ⅰ) Direct weighing: Weigh the mass of weighing bottle, bottle body and cap, and record the data in Table 1-1.	Electronic balance					Electric shock	Ensure cord is tagged	√	√			1	3	3	L	Report immediately to Demonstrator or Prep Room
Ⅱ) Weighing by difference: Weigh 0.4~0.5 g crude salt, 0.2~0.25 g crude salt and 0.10~0.13 g crude salt respectively. Record the data in Table 1-1.	Crude Salt	—	—	1 g		Toxic in contact with skin and if swallowed	Handle with care	√		√		1	3	3	L	Sponge & water; Report immediately to Demonstrator or Prep Room.
C: Data processing																	
Experimental Product (s)																
Experimental Waste (s)																
Final assessment of overall process	Low risk when correct procedures followed															
Signatures	Supervisor:				Lab Manager/OHS Representative:								Student/Operator:			

实验2　滴定分析的基本知识和操作

【实验目的】

1. 掌握酸碱滴定的原理。
2. 掌握滴定操作。
3. 学会正确判断滴定终点。

【实验原理】

滴定分析是最基本的定量分析之一,具有分析速度快、准确度高、应用范围广等特点。其原理是利用已知浓度的标准溶液与被测物质所发生的符合特定化学计量关系的定量反应,将该标准溶液装入滴定管作为滴定剂,滴加到被分析体系,测量出恰好与被测物质完全反应时所需滴定剂的体积,从而计算出被测物质的量。

【器材与试剂】

1. 器材:酸式滴定管,碱式滴定管,移液管,容量瓶,锥形瓶,烧杯。
2. 试剂:氢氧化钠,浓盐酸,酚酞指示剂,甲基橙指示剂。

● 视频演示

【实验内容】

1. 移液管和吸量管

为了准确地移取一定量体积的溶液,常用移液管(如图2-1a)和吸量管(如图2-1b)进行操作。其中移液管用于准确移取固定体积的溶液,而吸量管用于准确移取非固定体积的溶液。

在移液管和吸量管上标明的温度下,吸取溶液至弯月面与管刻度线相切,再让溶液按一定的方式自由流出,则流出溶液的体积与管上标明的体积相同。

移液前,先用自来水冲洗移液管和吸量管,若发现其内壁和外壁下部挂水珠,则须用洗涤液清洗。

在移取溶液前,用滤纸将管尖内外的水吸尽,以保证移取的溶液浓度不变。然后用被移取溶液将移液管或吸量管润洗2~3次。先将少量被移取溶液倒入干燥小烧杯中,将移液管或吸量管插入小烧杯中吸取溶液,当吸至移液管球部约1/4处,吸量管则至充满全部体积的1/4处,立即用右手食指按住管口,取出管子,缓慢放平并旋转,使液体将管内壁全部润湿。然后将溶液从管尖处弃去。反复操作2~3次。

移取溶液时,右手拇指和中指捏住管颈标线上方,将管尖插入待移取溶液

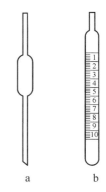

图2-1　移液管(a)和吸量管(b)

面下 1~2 cm 处,左手拿洗耳球,将洗耳球内空气排出,然后将洗耳球尖口紧接在管上口,慢慢松开左手手指,使待移取溶液吸入管内。移液管应随容器内液面的下降而下降,以免洗空。当液面上升至标线以上时,移去洗耳球,迅速用右手食指按紧管口。将移液管下口提至液面以上,将管下部贴在器壁上,除去管尖外壁溶液。将容器倾斜约 45°,竖直移液管,其管尖仍贴在器壁上,微微放松食指并用拇指和中指缓慢转动管子,液面缓慢下降,直至溶液的弯月面下缘与标线相切时,立即用食指按紧管口,使溶液不再流出。将移液管从容器中取出,移至倾斜约 45°的接收容器中,管尖紧靠接收容器内壁。松开食指,使溶液沿接收容器内壁流下。溶液流完后,停留 10~15 s 后,取出移液管,如图 2-2 所示。

除在管身标有"吹"字的管子外,不可将残留在管尖内壁处的少量溶液吹入接收容器中,因在校正移液管时,已将尖端内壁处的保留溶液体积考虑在内。

吸量管的使用方法同移液管类似,但其移取溶液的准确度不如移液管,使用吸量管时还应避免使用末端刻度。

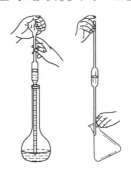

图 2-2 移液管的使用

2. 容量瓶及其使用方法

容量瓶主要用于配制准确浓度的溶液或将浓溶液稀释成准确浓度和体积的稀溶液。在指定温度下,溶液充满至弯液面下缘与容量瓶瓶颈标线相切时,所容纳的溶液体积等于瓶上标示的体积。

使用容量瓶前应先检查是否漏水。具体操作如下:向容量瓶加水至标线附近,塞紧瓶塞,左手持瓶颈标线以上位置按紧瓶塞,另一手扶住瓶底,将容量瓶倒置约 2 min,观察是否漏水,如果没有漏水,将瓶塞旋转 180°后再次检漏。两次检查均不漏水,方可使用。

容量瓶的洗涤与移液管相似。

由固体物质配制标准溶液,需先将精确称量的固体试剂放入小烧杯,加适量纯水,搅拌使其溶解后再通过玻璃棒引流将溶液定量转移至干净的容量瓶中,然后用洗瓶吹洗烧杯 3~4 次,吹洗液全部按上法转入容量瓶中。加水稀释至容量瓶容积 2/3 时,将容量瓶沿水平方向摇转几次(勿倒转),使溶液初步混匀。然后,将容量瓶平放在桌子上,继续加水至距标线以下约 1 cm,等待 1~2 min,使附着在瓶颈内壁的溶液充分流下。最后用滴管逐滴加水至溶液弯液面下缘与标线相切。盖好瓶塞,左手持瓶颈标线以上位置,食指按紧瓶塞,右手托住瓶底,倒转容量瓶并摇动,使瓶内气泡上升至顶部。如此反复数次,使溶液充分混匀。如图 2-3 所示。

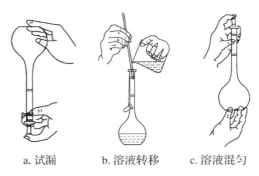

a. 试漏　　b. 溶液转移　　c. 溶液混匀

图 2-3 容量瓶的使用

如用容量瓶稀释溶液,用移液管(或吸量管)移取一定体积的浓溶液于容量瓶中,按上述方法加水稀释至标线,摇匀,即得。

3. 滴定管的准备与使用

滴定管一般分为酸式滴定管和碱式滴定管(如图 2-4 中 a 和 b)两种。

(1) 滴定管使用前的准备

酸式滴定管使用前应检查滴定管旋塞转动是否灵活,检查是否漏液。检漏具体操作如下:在滴定管内充满水,关闭旋塞,将滴定管夹在滴定管架上,静置 2 min,观察管口及旋塞处是否有漏液;将旋塞旋转 180°再放置 2 min,检查是否有漏液。若前后两次均无漏液,且旋塞转动灵活,则滴定管可正常使用,否则应重新涂油脂(凡士林或真空油脂)。

涂油脂方法如下:将滴定管中的水倒掉,平放在实验台上,抽出旋塞,用滤纸将旋塞及旋塞槽内的水吸干,用手指蘸少许油脂在旋塞的两头(如图 2-5 中 A 和 B),均匀地涂上薄薄的一层。将旋塞插入旋塞槽中,向同一方向旋转,直至旋塞除旋塞孔一圈外全部呈透明状。油脂不可涂得太多,以免堵塞旋塞孔;亦不可涂得太少,以免旋塞转动不灵活,甚至漏液。如遇到这些情况,均须将旋塞和旋塞槽擦干净后,重新涂油脂。最后,顶住旋塞大端,在另一端套上橡皮圈。

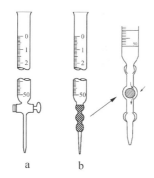
图 2-4 酸式滴定管(a)和碱式滴定管(b)

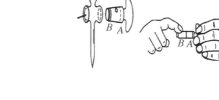
图 2-5 涂凡士林

滴定管应在进行完涂油脂检漏后,再充分清洗,直到管内壁润湿均匀形成水膜但不挂水珠。洗涤时,根据管内污染程度,可分别选用自来水、洗涤剂等清洁物品加以洗涤。

使用碱式滴定管前,应选择大小合适的玻璃珠和乳胶管,否则容易漏水或操作不便。

碱式滴定管的洗涤方法基本上与酸式滴定管相同。

(2) 标准溶液的装入

将标准溶液倒入滴定管前,应先将其摇匀,使凝结在瓶壁上的水珠混入溶液。随后,左手持滴定管上端无刻度处,将混好的标准溶液直接缓缓倒入滴定管中。

为了避免装入后的标准溶液被稀释,应先用标准溶液润洗滴定管三次(每次约 5~10 mL)。润洗时,两手平端滴定管,缓慢转动,使标准溶液润湿滴定管整个内壁。对于酸式滴定管,打开旋塞,让溶液冲下,冲洗出口管,放净残液。对于碱式滴定管,则弯曲下端乳胶管,使出口管管尖向上翘起,挤宽玻璃珠一侧的乳胶管,让溶液喷出,排出气泡(如图 2-6)。润洗结束后,将标准溶液倒入滴定管至"0"刻度以上,再调节液面在 0.00 mL 处或稍微靠下一点的位置。

图 2-6 碱式滴定管排出气泡

(3) 滴定管读数

滴定管读数时,应遵守以下原则:

① 装满溶液或放出溶液后,须静置 1~2 min 后再读数。如果放出溶液速度较慢(如临近滴定终点时),静置 30~60 s 后即可读数。

② 读数应在内壁不挂水珠、管尖无气泡的情况下完成。

③ 读数时,滴定管应从滴定管架上取下,用拇指和食指捏住管上端无刻度处,将滴定管保持垂直,进行读数。

④ 液体由于表面张力,滴定管内液面呈弯月形。滴定管内装入无色或浅色溶液时,读数视线应与液面水平,读取视线与弯月面下缘实线最低点相切处的刻度,如图 2-7 所示。对于深色溶液且看不清弯月面下缘时,可读取视线与液面两侧的最高点呈水平处的刻度。

⑤ 读数必须读至小数点后第二位,且要求准确到 0.01 mL。

⑥ 为了读数准确,可采用一张中部有黑色长方块的白色读数卡。读数时,将读数卡放在滴定管背面,使黑色部分在弯月面下约 1 mm 处,此时即可看到弯月面的反射呈黑色,读与此黑色弯月面下缘相切的刻度。读取深色溶液液面两侧最高点时,应衬以白色卡片为背景。读数时应采用同一方法。

⑦ 对于乳白板蓝线衬背的滴定管,应读取液面处蓝线形成的上下两个尖端和交点所对应的刻度值。

⑧ 读取初读数时,应将滴定管管尖悬挂的液滴除去。读取终读数时,管尖也不应挂液。如管尖挂有液滴,则此液滴已计入滴定体积之内,应使之进入被滴溶液中。若滴入后过量,则须重做。

(4) 滴定操作及方法

滴定时,应将滴定管垂直地夹在滴定管架上进行滴定。

使用酸式滴定管时,用左手控制滴定管的旋塞。左手拇指在前,食指和中指在后,无名指及小手指向手心弯曲,轻轻贴着出口管。转动时不要用力向外推,掌心不能顶旋塞小端,否则易导致旋塞松动,造成漏液;也不要过分往里用力,以防旋塞无法灵活转动。滴定操作如图 2-8 所示。

图 2-7 滴定管读数　　　　　　图 2-8 酸式滴定管滴定操作示意图

使用碱式滴定管时,左手拇指在前,食指在后,其余三指夹住出口管。用拇指与食指的指尖捏挤玻璃珠一侧的乳胶管,使乳胶管与玻璃珠之间形成一小缝隙,溶液由缝隙处流出。应当注意,不要用力捏玻璃珠,玻璃珠不可上下移动;不要捏玻璃珠下部的乳胶管,以免空气进入形成气泡;停止滴定时,应先松开拇指和食指,然后再松开其余三指。

滴定最好在锥形瓶中进行,必要时也可以在烧杯中进行。滴定时实验台应为白色背景,以便观察滴定过程中溶液颜色的变化。

使用锥形瓶进行滴定时,用右手拇指、食指和中指捏住瓶颈,锥形瓶瓶底离滴定台高约 2~3 cm。滴定管管尖伸入瓶口内约 1 cm。左手控制滴定速度,边滴加溶液,边用右手沿同一圆周方向旋摇锥形瓶,不可前后晃动,以防溶液溅出。

开始滴定时,可采用"连珠式"地连续滴加溶液(液流不能成线),左手始终维持控制旋塞,观察滴落点周围颜色的变化。临近终点时,只加入一滴、半滴甚至小半滴(让溶液在管尖上悬而未落)滴定液,将锥形瓶与管口接触,使液滴流出,并用洗瓶以纯水冲下,直到溶液颜色发生明显变化,停止滴加,滴定结束。

4. 配制 500 mL 0.10 mol·L⁻¹ NaOH 溶液

称取 2.0 g NaOH 于洁净的小烧杯中,加水 50 mL,全部溶解后转入 500 mL 试剂瓶中,将小烧杯用洗瓶中的少量水吹洗数次,将吹洗液全部转入试剂瓶中,再加水至总体积约 500 mL 左右,盖上橡皮塞,摇匀。

5. 配制 500 mL 0.10 mol·L⁻¹ HCl 溶液

在通风橱内用洁净的 10 mL 量筒量取 4.2~4.5 mL 浓盐酸,倒入盛有水的 500 mL 试剂瓶中,加水稀释至 500 mL 左右,盖上玻璃塞,摇匀。

6. 滴定操作练习

(1) 取酸式滴定管和碱式滴定管各一支,分别用 5~10 mL 配制好的 HCl 溶液和 NaOH 溶液润洗酸式滴定管和碱式滴定管 2~3 次。再在酸式滴定管和碱式滴定管中分别装入配制好的 HCl 溶液和 NaOH 溶液,排出气泡,调节液面至零刻度或稍下一点的位置,静止 1 min 后,记录初读数于表 2-1。

（2）用 NaOH 溶液滴定 HCl 溶液。取约 10 mL 配制好的 HCl 溶液于锥形瓶中,加 10 mL 水,滴入 1~2 滴酚酞作指示剂,用配制好的 NaOH 溶液进行滴定,直至终点,记录终读数于表 2-1。如此反复练习 3 次。

（3）用 HCl 溶液滴定 NaOH 溶液。取约 10 mL 配制好的 NaOH 于锥形瓶中,加 10 mL 水,滴入 1~2 滴甲基橙作指示剂,用配制好的 HCl 溶液进行滴定,直至终点,记录终读数于表 2-1。如此反复练习 3 次。

表 2-1 滴定分析基本操作数据处理

项目 \ 数据 序号	1	2	3
V_{HCl} 终读数/mL V_{HCl} 初读数/mL V_{HCl}/mL			
V_{NaOH} 终读数/mL V_{NaOH} 初读数/mL V_{NaOH}/mL V_{HCl}/V_{NaOH} V_{HCl}/V_{NaOH} 的平均值			
相对平均偏差/%			

【注意事项】

1. 移液管(吸量管)不能移取太冷或太热的液体;不能在烘箱里烘干;为避免误差,在同一实验中应使用同一移液管(吸量管)。

2. 容量瓶只能用于配制溶液,不能长期储存溶液;如果溶质在溶解过程中放热,则要待溶液冷却后再进行转移;需避光的溶液应以棕色容量瓶配制。

3. 酸式滴定管下端为玻璃活塞的不宜装入碱液。碱式滴定管内不宜装入氧化性溶液。

4. 洗涤碱式滴定管过程中,应注意玻璃珠下方"死角"处的清洗,在挤宽乳胶管放出溶液时,应不断改变挤的方位,以确保玻璃珠周围的各个角落都能被洗到。

5. 滴定结束后,滴定管内剩余的溶液应弃去,切勿倒回原瓶。

【思考题】

1. 用容量瓶配制溶液时,是否需要用待测溶液润洗? 为什么?
2. HCl 和 NaOH 溶液能直接配制准确浓度吗? 为什么?

风险评估及控制工作表

实验任务	滴定分析的基本知识和操作				评估人	吴泽颖，王红松，吴星星			评估日期	2021年7月1日		
过程步骤	危害鉴别				健康风险	安全措施			清单/风险评估	风险评估		措施
	化合物名称	类别	浓度	用量	化学品安全技术说明书	可能的伤害种类	降低或消除风险	技术 防护眼镜 手套 通风橱		可能性 1~4	危害 1~4	风险	危害级别 高中低	紧急情况/泄露	特殊备注

过程步骤	化合物名称	类别	浓度	用量	化学品安全技术说明书	可能的伤害种类	降低或消除风险	技术	防护眼镜	手套	通风橱	可能性	危害	风险	危害级别	紧急情况/泄露	特殊备注
					或物理危害												
滴定操作练习																	
I) 准备好酸式和碱式滴定管各一支。分别用5~10 mL HCl 和 NaOH 溶液润洗酸式和碱式滴定管2~3次。再在酸式滴定管和碱式滴定管中分别装入 HCl 和 NaOH 溶液，排出气泡，调节液面至零刻度或稍下一点的位置，静止1 min后，记录初读数于表2-1。	NaOH	8	0.1 M	500 mL	√	皮肤灼伤和眼损伤	关注		√	√		1	3	3	低	大量冲洗	
	HCl	2,3+8	0.1 M	500 mL	√	皮肤灼伤和眼损伤	关注		√	√		1	3	3	低	大量冲洗	
II) 用 NaOH 溶液滴定 HCl 溶液。取约10 mL 配制好的 HCl 溶液于锥形瓶中，加10 mL 水，滴入1~2滴酚酞作指示剂，用配制好的 NaOH 溶液进行滴定，直至终点，记录读数于表2-1。如此反复练习3次。	酚酞指示剂	—	1%	1 mL	—	—						1	1	1	低	大量冲洗	
III) 用 HCl 溶液滴定 NaOH 溶液。取约10 mL 配制好的 NaOH 溶液于锥形瓶中，加10 mL 水，滴入1~2滴甲基橙作指示剂，用配制好的 HCl 溶液进行滴定，直至终点，记录读数于表2-1。如此反复练习3次。	甲基橙指示剂	—	0.5%	1 mL	—	—						1	1	1	低	大量冲洗	
实验产物																	
实验废弃物																	
整个过程的最终评估	低风险（按照正确流程操作）																

签名　　实验老师：　　　　　实验室主任/职业安全健康（OHS）代表：　　　　　学生/操作人：

Experiment 2 Elementary knowledge and basic operation of titration analysis

【Objectives】

1. To understand the principles of acid-base titration.
2. To learn the titration procedure.
3. To properly determine the endpoint.

【Principles】

Titrimetric analysis is one of the most basic quantitative analysis, which is fast, accurate and widely used. In general, the standard solution of known concentration is put into burette as titrant, which is dripped into the analyte system to have quantitative reaction with the analyte according to the specific stoichiometric relationship. Measure the volume of titrant required for the complete reaction with the tested substance, and then calculate the quantity of the tested substance.

【Apparatus and Chemicals】

1. Apparatus

Acid burette; Alkali burette; Pipette; Volumetric flask.

2. Chemicals

NaOH; HCl; phenolphthalein indicator; methyl orange indicator.

【Experimental Procedures】

1. Pipette and graduated tube

In order to accurately move a certain volume of solution from one container to another, pipette (Figure 2 – 1 a) and graduated tube (Figure 2 – 1 b) are often used. The pipette is used to accurately transfer a fixed volume of solution, while the graduated tube is used to accurately transfer a non fixed volume of solution.

At the temperature indicated on the pipette and graduated tube, suck the solution and adjust the meniscus liquid level of the solution to be tangent to the scale line of the pipe, and discharge the solution in the specified way. The volume of the solution discharged is the same as that indicated on the pipe. If

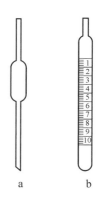

Figure 2 – 1 Pipette (a) and graduated tube (b)

the actual transfer temperature is different from the indicated temperature, or when the non-aqueous solvent is absorbed, the volume of the solution will be slightly different. Pipette and graduated tube can be corrected if necessary.

When the pipette and graduated tube are used, the right hand holds the part above the scale line of the pipette, the tip of the pipette is inserted into the solution to be sucked, and the left hand controls the rubber suction bulb, so that the rubber suction bulb is closely connected with the upper mouth of the pipette for suction. The insertion depth of the pipe tip shall be appropriate, generally 1~2 cm below the liquid level. The tip of the tube should drop as the liquid level drops during suction.

Before taking the solution, wash the pipette and graduated tube with tap water. If water drops are found on the inner wall and the lower part of the outer wall, wash them with detergent. First, remove the residual water, suck the lotion to about 1/4 of the ball part of the pipette, or 1/4 of the volume of the graduated tube. Remove the rubber suction bulb, press the right index finger tightly on the nozzle, and remove the lotion. Lay the tube flat, hold the middle of the tube with left hand, release the right index finger, rotate the tube, and let the inner wall of the tube be moistened by the lotion. Pour the lotion back to the original bottle from the top of the tube, wash it thoroughly with tap water, and then wash it with pure water three times. Every time the water is absorbed to about 1/4 of the ball part of the pipette, lay the tube flat and rotate to wash the inner wall, and then remove it. The lower part of the outer wall of the pipe should also be blown clean with pure water.

Before transferring the solution, use the filter paper to absorb all the water inside and outside the pipe tip, so as to ensure the constant solution concentration. Then, moisten the pipette with the solution for 2~3 times. Moistening method: pour a small amount of transferred solution into a small dry beaker, insert the tube into the beaker to absorb the solution, when it is sucked to about 1/4 of the ball part of the pipette, or filled with 1/4 of the total volume of graduated tube, immediately press the nozzle with the right index finger, take out the tube, slowly lay the tube flat and rotate, so that the liquid will wet the whole tube wall. Then discard the solution from the tip of the tube. Repeat it for 2~3 times.

When transferring the solution, hold the upper part of the neck line with the right thumb and middle finger, insert the lower part of the pipe to 1~2 cm below the liquid level, and hold the rubber suction bulb with the left hand in the form of a clenched fist. First, discharge the air in the rubber suction bulb, and then place the rubber suction bulb tip on the upper mouth of the pipe, slowly release the left finger, so that the solution to be transferred can be inhaled into the pipe. When the liquid level rises above the mark, remove the rubber suction bulb and quickly plug the nozzle with the right index finger. Lift the lower mouth of the tube to above the liquid level, but still lean on the wall of the device, slightly loosen the index finger, slowly rotate the tube with the thumb and middle finger.

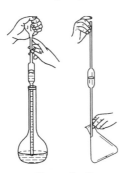

Figure 2 - 2
Usage of pipette

When the liquid level slowly drops to the bottom of the meniscus tangent to the marking line, remove the rubber suction bulb, immediately press the tube mouth with the index finger to make the solution no longer flow out. Take the pipette out of the container, use the filter paper to dry the solution on the outer wall of its lower end, but do not touch the lower mouth. Take the container with the left hand, incline about 45°, keep the pipette vertical and move it into the container with the right hand, and the lower end of the pipe is close to the inner wall of the container. Release the index finger so that the solution flows down the inner wall of the receiving container. After the solution flow is completed, stay for 10~15 s, and then take out the pipette. As shown in Figure 2 - 2.

Except for the pipe marked with "blow" on the pipe body, a small amount of solution remaining on the inner wall of the pipe tip should not be blown into the receiving container. The reason is that the volume of the solution remaining on the inner wall of the tip has been taken into account when calibrating the pipette.

After use, wash the pipette and place it on the pipette holder.

Graduated tubes are used in the same way as pipettes, but when the solution is discharged, it should be lowered from one scale to another and the end scale should be avoided.

2. Volumetric flask and its usage

A volumetric flask is a pear shaped, flat bottomed glass bottle with a ground glass or plastic stopper. There is a mark on its neck. At the specified temperature, when the solution is filled to the bottom of the meniscus tangent to the mark, the volume of the solution contained is equal to the volume marked on the bottle. The volumes of commonly used volumetric flask are 10 mL, 25 mL, 50 mL, 100 mL, 250 mL, 500 mL, 1 000 mL, etc. The main purpose of volumetric flask is to prepare accurate concentration solution, including preparing standard solution by direct method or diluting accurate concentration and volume of concentrated solution to accurate concentration and volume of dilute solution.

Check for water leakage before using the volumetric flask. The specific operation is as follows: add water to the volumetric flask near the marking line, and tighten the cork. One hand holds the cork tightly above the marking line of the bottle neck, and the other hand holds the bottom of the bottle. Invert the volumetric flask for about 2 min, and observe whether there is water leakage. If there is no water leakage, rotate the cork for 180° and check again. It can be used only when there is no water leakage in both inspections.

The washing of the volumetric flask is similar to that of the pipette.

If the standard solution is prepared with solid substances, put the accurately weighed solid reagent into a small beaker, add some pure water, and stir to make it dissolve (if it is difficult to dissolve, cover the surface dish and heat it slightly, but it must be cooled before transferring). Quantitatively transfer the solution to the volumetric flask with the glass rod, and then blow the wall of the beaker with the washing bottle for 3 to 4 times, and transfer the washing solution into the volumetric flask according to the same method. Add distilled water to dilute. When diluting to 2/3 of the volume of the volumetric flask, shake the volumetric flask several times in the horizontal direction (do not reverse), so as to preliminarily mix the solution. Then, place the volumetric flask flat on the table, continue to add distilled water to 1~2 cm away from the mark line, wait for 1~2 min, so that the solution attached to the inner wall of the bottleneck flows down. Finally, add distilled water drop by drop with a rubber head dropper until the bottom of the solution meniscus is tangent to the marking line. Close the cork, press the cork with the index finger of left hand, hold the bottom of the bottle with the fingers of right hand, turn the volumetric flask upside down, and make the bubble in the bottle rise to the top. Turn it over and over several times so that the solution is well mixed. As shown in Figure 2-3.

If diluting the solution with a volumetric flask, use a pipette (graduated tube) to transfer a certain volume of concentrated solution into the volumetric flask, add water according to the above method to dilute to the marking line, and shake well.

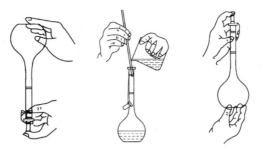

a. Leak test b. Solution transfer c. Solution mixing

Figure 2-3 Usage of volumetric flask

3. Preparation and operation of burette

There are two types of burette: acid burette and alkali burette (Figure 2-4 a and b). The commonly used burette solvent is 25 mL and 50 mL. The universal burette with polytetrafluoroethylene piston has been widely used.

（1）Preparation of burette before use

Before use, check whether the burette cock rotates flexibly and whether the burette cock matches with the plug groove, otherwise it is easy to cause leakage. For the matching burette, in order to make the piston rotate flexibly without leakage, oil (Vaseline or vacuum grease) can be applied to the part of the cock, and then leak detection. The operation is as follows.

Pour out the water in the burette, lay it on the test bench, draw out the cock, use the filter paper to suck up the water in the cock and the cock groove, dip a little grease in the finger, evenly apply a thin layer on both ends of the cock (Figure 2-5 A and B). Insert the plug into the plug groove and rotate in the same direction until the plug is transparent except for one turn of the plug hole. Grease shall not be applied too much, otherwise it is easy to block the plug hole. However, if it is applied too little, the plug will not rotate flexibly and even leak liquid. In case of these situations, the plug and plug groove must be wiped clean and then greased again. Finally, counter the large end of the cock and put a rubber band on the other end.

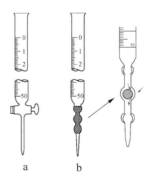
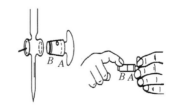

Figure 2-4 Acid burette (a) and alkali burette (b)   **Figure 2-5 Apply grease**

During leak detection, fill the burette with water, discharge the air bubble in the outlet pipe, close the cock, clamp the burette on the burette frame, and stand upright for 2 min. If there is no water drop leaking, rotate the cock 180° for static inspection. If water leakage is found, grease shall be applied again.

After the burette is coated with grease for leakage detection, it shall be fully cleaned to ensure that the inner wall of the burette is evenly wetted by water to form a water film without water drops. When washing, according to the degree of contamination in the pipe, tap water or detergent can be used respectively.

The lower end of the Alkali burette is connected with a latex tube embedded with glass beads to control the flow rate. The Alkali burette should not be filled with oxidizing solution.

Before using the Alkali burette, check whether the latex tube is aging and whether the glass bead size is appropriate, otherwise it should be replaced.

The washing method of Alkali burette is basically the same as that of acid burette.

(2) Loading of standard solution

The prepared standard solution shall be directly poured into the burette without using other containers (such as beaker, funnel, etc.) for transfer. Before pouring into the solution, shake the solution in the reagent bottle well, so that the water drops condensed on the bottle wall are mixed into the solution. Then, hold the top without scale of the burette with the first three fingers of the left hand, tilt the body of the burette slightly or make the burette naturally vertical, take the reagent bottle (label up) with the right hand and pour it slowly.

First, moisten the inner wall of the burette with solution for three times (about 5~10 mL each time). Pay attention to wash all the inner walls every time to ensure that the concentration of the solution is constant after loading. For acid burette, open the cock, wash the outlet pipe and drain the residual liquid.

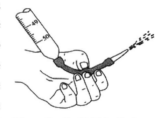

Figure 2-6 Bubble discharge from Alkali burette

Hold the top without scale of the burette with the right hand, open the cock quickly with the left hand, let the solution wash down, and drive out the bubbles. For the Alkali burette, bend the lower latex tube to make the outlet tube tip up, extrude the latex tube on one side of the glass bead to let the solution spray out, and bring out bubbles (Figure 2-6). After the bubbles are exhausted, the solution is discharged and the latex tube is straightened at the same time. Finally, wipe the outer wall of the burette dry and clamp it on the burette frame for use.

(3) Burette reading

The following principles should be observed.

① After the solution is put in or discharged, it must wait for 1~2 min, and read after the solution attached to the inner wall of the pipe flows down fully. If the solution is discharged slowly (for example, only one drop or half a drop of solution is added at a time near the end of titration), it can be read only after 30~60 s.

② Before each reading, check whether there are water drops on the inner wall of burette and whether there are bubbles on the tip of burette. The reading is valid only when there is no water drop on the inner wall and no bubble on the pipe tip.

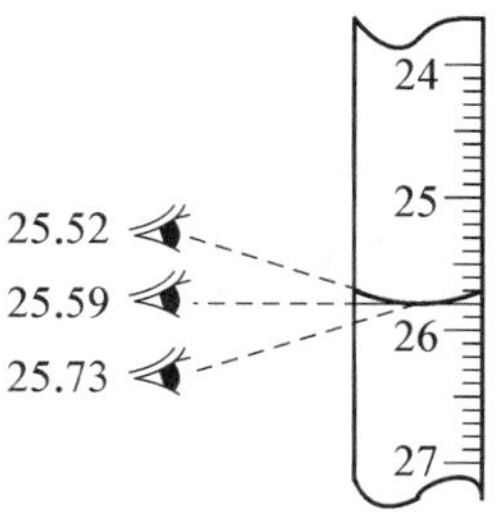

③ When reading, the burette can be clamped on the burette holder, or it can be removed. Hold the top without scale of the burette for reading, but pay attention to keep the burette vertical.

④ When reading, the line of sight should be level with the liquid level, otherwise the reading will be large or small (Figure 2-7). When the burette is filled with colorless or light color solution, the scale at the intersection of the line of sight and the lowest point of the meniscus liquid

Figure 2-7 Burette reading

level should be read. For dark solution, when the lower edge of meniscus cannot be seen clearly, the readable line of sight is horizontal with the highest point on both sides of meniscus.

⑤ The reading must be read to the second decimal place, i.e. 0.01 mL.

⑥ To facilitate reading, a white reading card with a black rectangle in the middle can be used. When reading, press the reading card against the back of the burette, so that the upper edge of the black square is about 1 mm below the meniscus liquid level. At this time, black is reflected on the meniscus, which is clear and easy to distinguish. The reading is based on the lowest point of the lower edge of the black meniscus. When reading the highest points on both sides of the dark solution liquid level, the white card should be used as the background. The same method should be used to read the readings.

⑦ For burette with blue line backing of opal board, the scale corresponding to the upper and lower tips and intersection points formed by blue line at liquid level should be read.

⑧ The initial reading should be adjusted at or near the "0" line. When taking the initial reading, remove the liquid drop hanging from the pipe tip at the same time. When the final reading is taken, the tip of the tube should not hang liquid. If there is a drop, the drop has been included in the titration volume, and it should be put into the solution to be dropped. If it is excessive after dripping, it must be redone.

(4) Titration

The burette should be clamped vertically on the burette holder for titration.

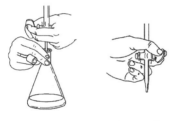

Figure 2 - 8　Titration used by acid burette

Use left hand to control cocks when using acid burette. The left ring finger and small finger point to the palm of the hand and bend, gently stick to the outlet pipe, and use the other three fingers to turn the cock. Do not push outwards when turning. The palm should not touch the small end of the cocks to prevent the cocks from being pushed loose and liquid leakage. Do not push too much inward so that the cocks cannot rotate flexibly. As shown in Figure 2 - 8.

When using the Alkali burette, the left ring finger and little finger clamp the glass outlet tube, and the thumb and index finger squeeze the latex tube on one side of the glass bead to make the solution flow out of the gap. In order to avoid leakage or air bubbles entering back, attention should be paid to: first, the glass bead should not be displaced up and down; second, do not pinch the latex tube under the glass bead; third, when stopping titration, loosen the thumb and index finger first, then the ring finger and small finger.

No matter which burette is used, the following three methods of dropping solution must be mastered: continuous dropping of solution (liquid flow cannot be linear); adding only one drop of solution; dropping half a drop or even half a drop, that is, let the solution hang on the tip of the tube without falling, contact the conical bottle with the mouth of the tube, make the drop flow out, and wash it down with pure water.

The titration is generally carried out in an Erlenmeyer flask, or in a beaker if necessary, and the test bench should be white background.

If an Erlenmeyer flask is used for titration, hold the bottle neck with the first three fingers of the right hand to make the bottom of the flask about 2 cm away from the table top. At the same time, adjust the height of burette so that the tip of burette is about 1 cm into the bottle mouth. Operate the burette with the left hand according to the above method, and add the solution while rotating the conical flask with the right hand.

The following points should be noted in titration:

(1) When rotating the conical flask, move the wrist joint slightly to make the solution move in a circle in the same direction (clockwise or anticlockwise), but do not shake back and forth to prevent splashing. The mouth of the bottle should not touch the tip of the burette when it is rotated.

(2) During titration, the left hand cannot leave the cock and let the solution to flow by itself.

(3) At the beginning of titration, the speed of adding solution can be slightly faster. During this period, pay attention to observe the color change of solution at the dropping point of the solution to determine whether it is close to the end point. When approaching the end point, the speed of adding solution should be slowed down. Add one drop, half drop or even a small half drop, shake it for several times after dropping, and blow the bottle wall with a little pure water. When the color of the solution changes obviously, close the cock quickly and stop titration, which is the end point of titration.

(4) When adding half a drop with an Alkali burette, it is necessary to avoid bubbles at the tip of the burette. After the extruded solution is hung on the tube tip, first release the thumb and index finger, then touch the liquid drop with the bottle wall, introduce it into the bottle, finally release the ring finger and small finger, and blow the bottle wall with a little pure water.

4. Preparation of 500 mL 0.10 mol·L^{-1} NaOH solution

Use a clean small beaker to weigh 2.0 g NaOH on the platform scale, add 50 mL pure water to make it dissolve completely, transfer it into a 500 mL reagent bottle, rinse the small beaker with a small amount of pure water for several times, transfer the washing solution into the reagent bottle together, add water to the total volume of about 500 mL, cover the rubber plug, and shake well.

5. Preparation of 500 mL 0.10 mol·L^{-1} HCl solution

Take 4.2~4.5 mL concentrated hydrochloric acid with a 10 mL clean measuring cylinder in the fume hood, pour it into a 500 mL reagent bottle containing water, add water to dilute it to about 500 mL, cover it with a glass stopper, and shake it well.

6. Titration practice

(1) Prepare one acid burette and one alkali burette. Rinse the acid burette and the basic burette with 5~10 mL HCl and NaOH solution for 2~3 times respectively. Add HCl and NaOH solution respectively, discharge bubbles, and adjust the liquid level to zero scale or slightly lower. Record the initial reading in Table 2-1 after standing for 1 min.

(2) Phenolphthalein was used as indicator to titrate HCl with NaOH solution. Drain about 10 mL HCl from the acid burette into a conical flask, add 10 mL pure water and 1~2 drops of phenolphthalein. Shake continuously. Titrate with NaOH solution until the end point, record the final reading in Table 2-1. Repeat this three times.

(3) Methyl orange was used as indicator to titrate NaOH with HCl solution. Drain about 10 mL NaOH from the basic burette into the conical flask, add 10 mL pure water and 1~2 drops of methyl orange. Shake continuously. Titrate with HCl solution until the end point, record the final reading in Table 2-1. Repeat this three times.

Table 2-1 Basic operation data processing of titration analysis

product \ Number data	1	2	3
V_{HCl} Final reading/mL V_{HCl} Initial reading/mL V_{HCl}/mL			
V_{NaOH} Final reading/mL V_{NaOH} Initial reading/mL V_{NaOH}/mL V_{HCl}/V_{NaOH} average value of V_{HCl}/V_{NaOH}			
RAD/%			

【Notes】

1. The pipette should not be used to transfer too cold or too hot liquid. It should not be dried in the oven. In order to avoid errors, the same pipette should be used in the same experiment.

2. The volumetric flask can only be used to prepare the solution, not to store the solution for a long time. If the solute is exothermic during dissolution, it should be transferred after the solution is cooled. The solution to be kept away from light should be prepared in a brown volumetric flask.

3. Acid burette with glass piston at the lower end should not be filled with alkali liquor.

4. When washing with water, pay attention to the cleaning of the corner under the glass bead. When extruding the latex tube to release the solution, change the extruding direction constantly so that the glass bead can be washed around.

5. After titration, the remaining solution in the burette should be discarded and never be returned to the original bottle.

【Questions】

1. Is it necessary to moisten the volumetric flask with solution to be tested when preparing the solution? Why?

2. Can HCl and NaOH solution directly prepare accurate concentration? Why?

第 2 章　实验基本操作

Risk Assessment and Control Worksheet

TASK	Assessor	Date
Elementary knowledge and basic operation of titration analysis	Zeying WU, Hongsong WANG, Xingxing WU	1 JUL. 2021

Process Steps	Hazard Identification				Health Risk	Safety Measures				Risk Assessment				Action	
	Chemical Name	Class	Concn.	Amount	MSDS	Type of Injury Possible	To reduce or eliminate the risk	Technique	Glasses	Gloves	Fume Cbd.	Checklist＼Risk Assessment	Risk Score	Risk Level	In case of Emergency/Spill	Special Comments
		or Physical Hazard										Likely	Hazard	H M L		
												1～4	1～4			

Titration practice

Process Steps	Chemical Name	Class	Concn.	Amount	MSDS	Type of Injury Possible	To reduce or eliminate the risk	Technique	Glasses	Gloves	Fume Cbd.	Likely	Hazard	Risk Score	Risk Level	In case of Emergency/Spill
Ⅰ) Prepare one acid burette and one alkali burette. Rinse the acid burette and the basic burette with 5～10 mL HCl and NaOH solution for 2～3 times respectively. Add HCl and NaOH solution respectively, discharge bubbles, and adjust the liquid level to zero scale or slightly lower. Record the initial reading in Table 2-1 after standing for 1 min.	NaOH	8	0.1 M	500 mL	√	Skin burns and eye injuries	Handle with care		√	√		1	3	3	L	Sponge & water
	HCl	2.3 +8	0.1 M	500 mL	√	Skin burns and eye injuries	Handle with care		√	√		1	3	3	L	Sponge & water
Ⅱ) Phenolphthalein was used as indicator to titrate HCl with NaOH solution. Drain about 10 mL HCl from the acid burette into a conical flask, add 10 mL pure water and 1～2 drops of phenolphthalein. Shake continuously. Titrate with NaOH solution until the end point, record the final reading in Table 2-1. Then put 1～2 mL HCl solution into the acid burette, continue to titrate with NaOH solution to the end point. Repeat this three times.	Phenolphthalein indicator	—	1%	1 mL	—							1	1	1	L	Sponge & water

continued

Risk Assessment and Control Worksheet

TASK	Elementary knowledge and basic operation of titration analysis				Assessor	Zeying WU, Hongsong WANG, Xingxing WU				Date	1 JUL. 2021		
Process Steps	Hazard Identification				Health Risk	Safety Measures				Risk Assessment			Action
	Chemical Name	Class	Concn.	Amount	Type of Injury Possible	To reduce or eliminate the risk	Technique	Glasses	Gloves	Fume Cbd.	Checklist＼Risk Assessment		In case of Emergency/Spill
				MSDS or Physical Hazard							Likely	Hazard	Risk Score	Risk Level H M L	Special Comments
											1~4	1~4		
Ⅲ) Methyl orange was used as indicator to titrate NaOH with HCl solution. Drain about 10 mL NaOH from the basic burette into the conical flask, add 10 mL pure water and 1~2 drops of methyl orange. Shake continuously. Titrate with HCl solution until the end point, record the final reading in Table 2-1. Then put 1~2 mL NaOH solution into the basic burette, continue to titrate with HCl solution to the end point. Repeat this three times.	Methyl orange indicator	—	0.5%	1 mL	—						1	1	1	L	Sponge & water

Experimental Product (s)

Experimental Waste (s)

Final assessment of overall process: Low risk when correct procedures followed

Signatures Supervisor: Lab Manager/OHS Representative: Student/Operator:

第 3 章
化学反应原理与物理量测定

实验 3　摩尔气体常数的测定

【实验目的】

1. 了解一种测定摩尔气体常数的方法。
2. 熟悉分压定律与气体状态方程的应用。
3. 练习分析天平的使用与测量气体体积的操作。

【实验原理】

根据气体状态方程式的表达式：

$$pV = nRT = m/M_r \cdot RT$$

式中：p——气体的压力或分压(Pa)

　　　V——气体体积(L)

　　　n——气体的物质的量(mol)

　　　m——气体的质量(g)

　　　M_r——气体的摩尔质量(g·mol^{-1})

　　　T——气体的温度(K)

　　　R——摩尔气体常数(文献值：8.31 Pa·m^3·K^{-1}·mol^{-1} 或 J·K^{-1}·mol^{-1})

可得，在一定的温度和压力条件下，若能测出一定量气体的体积，则可求得气体常数。

本实验采用铝与盐酸的置换反应，反应的化学方程式如下：

$$2Al(s) + 6HCl(aq) = 2AlCl_3(aq) + 3H_2(g)\uparrow$$

反应所生成的氢气可近似认为在实验条件下的理想气体，利用排水集气法收集并测量其体积，从而计算出气体常数。

由于氢气是在水面上收集的，其分压[$p(H_2)$]与水的饱和蒸气压[$p(H_2O)$]有关，根据分压定律：

$$p = p(H_2) + p(H_2O)$$

则 $p(H_2) = p - p(H_2O)$

式中 p 为大气压,可由气压计读出。

氢气的物质的量可根据铝的重量和原子量,由化学方程式计算出来。

由于 $p(H_2)$、$V(H_2)$、$n(H_2)$、T 均可由实验测得,这样根据气体状态方程式即可求得气体常数。

【器材与试剂】

1. 器材:分析天平,气压计,精密温度计,10 mL 量筒,产生和测定氢气体积的装置,铝片。
2. 试剂:6 mol·L^{-1} HCl。

【实验内容】

1. 铝片的称量

准确称取 15~25 mg 之间的一小片铝片(铝片不宜过重,以免产生的氢气的体积超过量气管的测量限度),记录铝片的重量于表 3-1。

2. 检查实验装置的气密性

实验装置如图 3-1 所示。在水准瓶中加入适量的水。将水准瓶微微上下移动,使水准瓶和量气管两个液面在同水平面上,且应在 0 刻度附近。塞紧支管试管上的胶塞。将水准瓶下移约在量气管 30 刻度左右,此时可见量气管的液面下降,但下降幅度有限。继续观察液面,待确认液面不再下降,则说明实验装置不漏气,可以继续操作。(若液面继续下降,甚至降到与水准瓶液面持平,说明实验装置漏气,须查找原因并改正后方能接续下面的操作)

3. 铝片及盐酸的装入

用长颈漏斗向试管底部注入 5 mL 6 mol·L^{-1} HCl,避免 HCl 粘到试管壁上。将铝片蘸少量水,用玻璃棒沿试管壁将其送入试管,使其粘到试管壁上。(注意:铝片不能与盐酸相接触。)

4. 氢气体积的测量

用胶塞塞紧试管固定在铁架台上,调节水准瓶高度,使之与量气管内的液面保持在同一水平面上,并稳定在量气管 0 刻度附近(必要时再检查一次实验装置是否漏气),记录此时的量气管液面准确刻度读数,记为 V_1,轻轻敲击试管使其中的铝片落入盐酸中,铝与盐酸反应生成氢气,产生的压力会使量气管的液面不断下降。(注意:随时将漏斗缓慢下移,以防止量气管中气体压力过高使气体漏出。)待反应停止,试管冷却到室温后,调节两个液面在同一高度,读取此时量气管液面准确刻度读数,记为 V_2。将数据记录于表 3-1。测量并记录实验温度 T 和大气压 p。将数据记录于表 3-1。

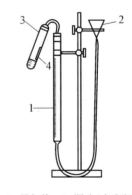

1. 量气管 2. 漏斗(水准瓶)
3. 试管(反应器) 4. 铝片

图 3-1 气体常数测定装置

表 3-1 摩尔气体常数的测定

检测项目	实验数据
铝片质量 m_{Al}/g	
反应前量气管内液面的读数 V_1/mL	
反应后量气管内液面的读数 V_2/mL	

续　表

检测项目	实验数据
反应置换出 H_2 的体积 $V=(V_2-V_1)\times 10^{-6}$/mL	
室温 T/K	
大气压力 p/Pa	
室温时水的饱和蒸气压 $p(H_2O)$/Pa	
氢气的分压 $p(H_2)=[p-p(H_2O)]$/Pa	
氢气的物质的量 $n(H_2)$/mol	
摩尔气体常数 $R=p(H_2)\cdot V/n(H_2)\cdot T$(J·K^{-1}·mol^{-1})	
相对误差 $=(R_{实验值}-R_{文献值})/R_{文献值}\times 100\%$	

【思考题】

1. 为什么必须检查实验装置是否漏气？实验中曾两次检查实验装置是否漏气，哪次相对更重要？

2. 在读取量气管液面刻度时，为什么要使漏斗和量气管两个液面在同一水平面上？

● 配套课件
● 知识点动画

风险评估及控制工作表

实验任务	摩尔气体常数的测定			评估人	吴泽颖,王红松,吴星星				评估日期	2021 年 7 月 1 日			
过程步骤	危害鉴别			健康风险		安全措施			评估清单/风险评估	风险评估			措施
	化合物名称	类别或物理危害	浓度	用量	化学品安全技术说明书	可能的伤害种类	降低或消除风险	技术	防护眼镜	手套	通风橱	可能性	危害	风险	危害级别	紧急情况/泼洒	特殊备注

过程步骤	化合物名称	类别或物理危害	浓度	用量	化学品安全技术说明书	可能的伤害种类	降低或消除风险	技术	防护眼镜	手套	通风橱	可能性	危害	风险	危害级别	紧急情况/泼洒	特殊备注
												1～4	1～4		高中低		
取下胶塞,用长颈漏斗向支管试管中加入 5 mL 6 mol·L^{-1} HCl,避免 HCl 粘到试管壁上。	盐酸	2,3 +8	6 M	5 mL	√	皮肤灼伤和眼损伤	关注		√	√	√	1	3	3	低	大量水冲洗	
实验产物																	
实验废弃物																	
整个过程的最终评估	低风险(按照正确流程操作)																
签名	实验老师:				实验室主任/职业安全健康(OHS)代表:							学生/操作人:					

Experiment 3　　Determination of molar gas constant

【Objectives】

1. To understand a method for determining the molar gas constant.
2. To be familiar with the application of partial pressure law and gas equation of state.
3. To practice using of analytical balances and operating of measuring gas volumes.

【Principles】

The expression of the gas state equation is:

$$pV = nRT = m/M_r \cdot RT$$

In the formula: p——gas pressure or partial pressure (Pa)

V——Gas volume(L)

n——the amount of gas substance(mol)

m——mass of gas(g)

M_r——The molar mass of the gas(g · mol^{-1})

T——gas temperature(K)

R——mole gas constant(document value: 8.31 Pa · m^3 · K^{-1} · mol^{-1} or J · K^{-1} · mol^{-1})

Therefore, for a certain amount of gas, if the volume occupied be measured under a certain temperature and pressure conditions, the gas constant can be obtained. In this experiment, aluminum is used to react with hydrochloric acid.

$$2Al(s) + 6HCl(aq) = 2AlCl_3(aq) + 3H_2(g) \uparrow$$

The generated hydrogen is approximately considered to be an ideal gas under experimental conditions, and then the volume is collected and measured by the drainage gas collection method, thereby the gas constant could be determined.

Since hydrogen is collected above the water surface, the partial pressure of hydrogen $[p(H_2)]$ is related to the saturated vapor pressure of water $[p(H_2O)]$. According to the law of partial pressure: $p = p(H_2) + p(H_2O)$, then $p(H_2) = p - p(H_2O)$ while p is atmospheric pressure and can be read by a barometer.

The amount of hydrogen can be calculated from the reaction formula based on the weight and atomic weight of aluminum.

Since $p(H_2)$, $V(H_2)$, $n(H_2)$ and T can be measured experimentally, the gas constant can be obtained according to the gas state equation.

【Apparatus and Chemicals】

1. Apparatus

Analytical balance; barometer; precision thermometer; graduated cylinder (10 mL); device for generating and measuring hydrogen volume; aluminum sheet.

2. Chemicals

6 mol·L^{-1} HCl.

【Experimental Procedure】

1. Weighing of aluminum flakes

Take a small piece of aluminum and weigh it on an electronic balance, and its weight must be between (20~30) mg (do not overweight the aluminum piece to avoid the volume of hydrogen generated exceeding the measurement limit of the gas measuring tube), and record the weight of the aluminum piece in Table 3-1.

2. Check the air tightness of the experimental device

The experimental setup is shown in figure 3-1. Add the suitable water to the level bottle. Move the level bottle up and down slightly so that the two liquid levels of the level bottle and the measuring tube are on the same level, and should be near the 0 scale. Tighten the rubber stopper on the test tube. Move the level bottle down to about 30 marks on the gas measuring tube, and the liquid level in the gas measuring tube will accordingly drops, but it will be stable after a short period of decline. Continue observing for a few minutes to confirm that the liquid level no longer drops, which indicate that the experimental device does not leak air, and then the following operations should be continue. (If the liquid level continues to drop, even down to the bottom of the level bottle, it means that the experimental device is leaking. You must find out the problem and fix it before the operation below).

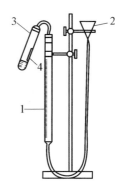

1. Gas measuring tube
2. Funnel (Level bottle)
3. Test tube (reactor)
4. Aluminum sheet

Figure 3-1 Gas constant measuring device

3. Loading of aluminum flakes and hydrochloric acid

Remove the rubber stopper and use a long-necked funnel to add 5 mL of 6 mol·L^{-1} HCl to the reaction tube to avoid HCl sticking to the wall of the test tube. Dip the aluminum flakes in a small amount of water and use a glass rod to send it along the wall into the test tube. (Note: the aluminum flakes cannot be in contact with hydrochloric acid.) Make it stick to the wall of the test tube. Record the data in Table 3-1.

4. Measurement of hydrogen volume

Connect the test tube to the rubber stopper tightly, adjust the height of the funnel to the same level as the liquid level of the gas measuring tube, and stabilize it near the 0 scale (if necessary, check the experimental device again for leaks), record the the gas pipe liquid level scale reads V_1. Gently shake the test tube (but do not remove it) to make the aluminum piece fall into hydrochloric acid. The liquid level of the gas pipe will drop continuously due to the pressure generated by the reaction between aluminum and hydrochloric acid. (Note: As the reaction happens, the funnel should be slowly moved down accordingly so that the liquid level in the gas measuring tube and the funnel are basically on the same plane to prevent the pressure in the gas measuring tube from being too high and causing gas leak). After the reaction is stopped, the test tube is cooled down to room temperature, and the two liquid levels are adjusted to be on the same horizontal surface, and then read the liquid level calibration as V_2. Record the data in Table 3-1. Measure and record the experimental temperature T and atmospheric pressure p. Record the data in Table 3-1.

Table 3-1 Determination of molar gas constant

Test items	Data recording
Quality of aluminum sheet m_{Al}/g	
The liquid level in eudiometer before reaction V_1/mL	
The liquid level in eudiometer after reaction V_2/mL	
Volume of hydrogen produced by reaction $V=(V_2-V_1)\times 10^{-6}$/mL	
Room temperature T/K	
Atmospheric pressure p/Pa	
Saturated vapor pressure of water under room temperature $p(H_2O)$/Pa	
Partial pressure of hydrogen $p(H_2)=[p-p(H_2O)]$/Pa	
Substantial amount of hydrogen $n(H_2)$/mol	
Molar gas constant $R=p(H_2)\cdot V/n(H_2)\cdot T(J\cdot K^{-1}\cdot mol^{-1})$	
Relative error $=(R_{experimental\ data}-R_{Literature\ data})/R_{Literature\ data}\times 100\%$	

【Questions】

1. Why is it necessary to check the experimental device for leaks? During the experiment, the device should be checked twice for leaks. Which one is more important?

2. When reading the liquid level of a gas measuring tube, why should the two liquid levels of the funnel and the gas measuring tube be on the same level?

Risk Assessment and Control Worksheet

TASK	Determination of molar gas constant	Assessor	Zeying WU, Hongsong WANG, Xingxing WU	Date	1 JUL. 2021

Process Steps	Hazard Identification					Health Risk		Safety Measures				Checklist — Risk Assessment			Action
	Chemical Name	Class	Concn.	Amount	MSDS	Type of Injury Possible	To reduce or eliminate the risk	Technique	Glasses	Gloves	Fume Cbd.	Likely	Hazard	Risk Score	Risk Level H M L	In case of Emergency/Spill
		or Physical Hazard											1~4	1~4			Special Comments
Remove the rubber stopper and use a long-necked funnel to add 5 mL of 6 M HCl to the reaction tube to avoid HCl sticking to the wall of the test tube.	HCl	2.3 +8	6 M	5 mL	√	Skin burns and eye injuries	Handle with care		√	√	√	1	3	3	L	Sponge & water
Experimental Product (s)																
Experimental Waste (s)																
Final assessment of overall process	Low risk when correct procedures followed																

Signatures	Supervisor:	Lab Manager/OHS Representative:	Student/Operator:

实验 4　化学反应速率和活化能

【实验目的】

1. 了解浓度、温度及催化剂对化学反应速率的影响。
2. 测定过二硫酸铵与碘化钾反应时的速率。
3. 学会计算反应级数、反应速率常数及反应的活化能。

【实验原理】

在水溶液中过二硫酸铵和碘化钾发生如下反应：

$$S_2O_8^{2-} + 3I^- =\!=\!= 2SO_4^{2-} + I_3^- \tag{4-1}$$

其反应速率方程可表示为：

$$v = k \cdot C^m(S_2O_8^{2-}) \cdot C^n(I^-)$$

式中，v 为反应速率；k 为速率常数；$C(S_2O_8^{2-})$、$C(I^-)$ 为即时浓度；m 与 n 之和称为反应级数。

通过实验能测定在单位时间内反应的平均速率（$\bar{v}$），如果在一定时间 Δt 内 $S_2O_8^{2-}$ 的浓度变化为 $\Delta C(S_2O_8^{2-})$，则平均速率可表示为：

$$\bar{v} = \Delta C(S_2O_8^{2-})/\Delta t$$

当 $\Delta t \to 0$ 时，可用平均速率代替瞬时速率。即：

$$\Delta C(S_2O_8^{2-})/\Delta t \approx v = k \cdot C^m(S_2O_8^{2-}) \cdot C^n(I^-)$$

为了测定在一定时间 Δt 内 $S_2O_8^{2-}$ 的浓度变化值，在将 $(NH_4)_2S_2O_8$ 和 KI 混合的同时，加入定量的 $Na_2S_2O_3$ 溶液和淀粉指示剂，这样在反应(4-1)进行的同时，体系中还进行着如下反应：

$$2S_2O_3^{2-} + I_3^- =\!=\!= S_4O_6^{2-} + 3I^- \tag{4-2}$$

反应(4-2)进行得非常快，瞬间即可完成，而反应(4-1)比反应(4-2)慢得多，由反应(4-1)生成的 I_3^- 立即与 $S_2O_3^{2-}$ 反应，生成无色的 $S_4O_6^{2-}$ 和 I^-。因此，在反应刚开始的一段时间内看不到 I_3^- 与淀粉相遇所呈现的特有蓝色。一旦 $Na_2S_2O_3$ 耗尽，反应(4-1)继续生成的 I_3^- 很快与淀粉作用而显现蓝色。

从反应(4-1)和(4-2)可以看出，$\Delta C(S_2O_8^{2-})$ 等于 $\Delta C(S_2O_3^{2-})$ 的 1/2。

在本实验中，每份混合液中 $Na_2S_2O_3$ 的起始浓度都相同。由于溶液呈现蓝色标志着 $S_2O_3^{2-}$ 全部耗尽，故从 $Na_2S_2O_3$ 的起始浓度可求出 $\Delta C(S_2O_3^{2-})$，进而可计算 $\Delta C(S_2O_8^{2-})$ 和反应的平均速率。

对反应速率方程两边取对数得：

$$\lg v = m \lg C(S_2O_8^{2-}) + n \lg C(I^-) + \lg k$$

当控制 $C(I^-)$ 不变时，以 $\lg v$ 对 $\lg C(S_2O_8^{2-})$ 作图，可得一直线，斜率为 m。同理，当 $C(S_2O_8^{2-})$ 不

变时，以 $\lg v$ 对 $\lg C(I^-)$ 作图，可求出 n。

求出 m 和 n 后，可用一组 $C(S_2O_8^{2-})$、$C(I^-)$ 和 v 数据代入速率方程求得反应速率常数 k。

根据阿伦尼乌斯公式，反应速率常数 k 与反应温度 T 之间有以下关系：

$$\lg\{k\} = -(E_a/2.303RT) + A$$

式中，E_a 为反应的活化能，$kJ \cdot mol^{-1}$；R 为摩尔气体常数，$R = 8.314 \ J \cdot K^{-1} mol^{-1}$；$T$ 为热力学温度；A 为常数项。

测出不同温度时的 k 值，以 $\lg k$ 对 $1/T$ 作图，可得一直线，由直线斜率 $\left(\dfrac{-E_a}{2.303R}\right)$ 可求得反应的活化能 E_a。

【器材与试剂】

1. 器材：100 mL 烧杯，大试管，量筒，电热恒温水浴槽，秒表，温度计。
2. 试剂：$(NH_4)_2S_2O_8$（0.2 $mol \cdot L^{-1}$，新配制），KI（0.2 $mol \cdot L^{-1}$），KNO_3（0.2 $mol \cdot L^{-1}$），$Na_2S_2O_3$（0.05 $mol \cdot L^{-1}$），$(NH_4)_2SO_4$（0.2 $mol \cdot L^{-1}$），$Cu(NO_3)_2$（0.02 $mol \cdot L^{-1}$），淀粉溶液（0.2%）。

【实验内容】

1. 浓度对化学反应速率的影响

在室温下，用量筒准确量取表 4-1 所列用量的各试剂，除 $(NH_4)_2S_2O_8$ 溶液外，其余各试剂均可按用量混合在对应编号的烧杯中，当加入 $(NH_4)_2S_2O_8$ 溶液时，立即计时，将溶液混合均匀，待溶液变为蓝色时，停止计时，记录时间 Δt 及室内温度。

表 4-1 浓度对化学速率的影响　　　　　　　　　室温：_____℃

	实验编号	1	2	3	4	5
V/mL	$(NH_4)_2S_2O_8$（0.2 $mol \cdot L^{-1}$）	10	5	2.5	10	10
	KI（0.2 $mol \cdot L^{-1}$）	10	10	10	5	2.5
	$Na_2S_2O_3$（0.05 $mol \cdot L^{-1}$）	3	3	3	3	3
	KNO_3（0.2 $mol \cdot L^{-1}$）				5	7.5
	$(NH_4)_2SO_4$（0.2 $mol \cdot L^{-1}$）		5	7.5		
	淀粉溶液（0.2%）	1	1	1	1	1
$C_0(S_2O_8^{2-})/(mol \cdot L^{-1})$						
$C_0(I^-)/(mol \cdot L^{-1})$						
$C_0(S_2O_3^{2-})/(mol \cdot L^{-1})$						
$\Delta t/s$						
$\Delta(S_2O_3^{2-})/(mol \cdot L^{-1})$						
$v/(mol \cdot L^{-1} \cdot s^{-1})$						
$k/[(mol \cdot L^{-1})^{1-m-n} \cdot s^{-1}]$						

2. 温度对化学反应速率的影响

按表 4-1 中实验 1 的试剂用量分别在高于室温 5 ℃、10 ℃ 和 15 ℃ 的温度下继续进行实验。将测得的各温度下的反应时间、计算得出的各温度下的反应速率及速率常数记录于表 4-2。

表 4-2 温度对化学反应速率的影响

实验编号	T/K	$\Delta t/s$	$v/(\mathrm{mol \cdot L^{-1} \cdot s^{-1}})$	$k/[(\mathrm{mol \cdot L^{-1}})^{1-m-n} \cdot \mathrm{s^{-1}}]$	$\lg\{k\}$
1					
6					
7					
8					

根据表 4-2 中的数据，以 $\lg k$ 对 $1/T$ 作图，求出直线的斜率，进而求出反应 (4-1) 的活化能。

3. 催化剂对化学反应速率的影响

在室温下，按表 4-1 中实验 1 的试剂用量，再分别加入 1 滴、5 滴、10 滴 $Cu(NO_3)_2$ 溶液，不足 10 滴的实验迅速用 $(NH_4)_2S_2O_8$ 溶液补充至 10 滴，搅拌，计时，把此时实验的反应速率与表 4-1 中实验 1 的反应速率比较。

表 4-3 催化剂对反应速率的影响

实验编号	9	10	11
加入 $Cu(NO_3)_2$ 溶液的滴数			
$\Delta t/s$			
$v/(\mathrm{mol \cdot L^{-1} \cdot s^{-1}})$			

【思考题】

1. 实验中为什么可以由反应溶液出现蓝色的时间长短来计算反应速率？反应溶液出现蓝色后，反应是否就终止了？
2. 试分析本实验中 $Na_2S_2O_3$ 的用量过多或过少，对实验结果有什么影响？

● 配套课件

风险评估及控制工作表

实验任务	化学反应速率和活化能					评估人	吴泽颖、王红松、吴星星			评估日期	2021年7月1日				
过程步骤	危害鉴别			化学品安全技术说明书	健康风险		降低或消除风险	安全措施			清单/风险评估	风险评估			紧急情况/泼洒措施
	化合物名称	类别	浓度 用量		可能的伤害种类			技术	防护眼镜	手套 通风橱		可能性	危害	风险	危害级别
		戒物理危害										$1\sim4$	$1\sim4$		高中低	特殊备注

A：浓度对化学反应速率的影响，求反应级数

过程步骤	化合物	类别	浓度	用量	SDS	伤害种类	降低风险	技术	眼镜	手套	通风橱	可能性	危害	风险	级别	措施
在室温下，用量筒准确量取表4-1所列用量的各试剂，除$(NH_4)_2S_2O_8$用量在对应编号的烧杯中，当加入$(NH_4)_2S_2O_8$溶液时，立即计时，将溶液混合均匀，待溶液变为蓝色时，停止计时，记录时间Δt及室内温度。	$(NH_4)_2S_2O_8$	5.1	0.2 M	10 mL	√	皮肤刺激	关注			√		1	3	3	低	大量冲洗
	KI	—	0.2 M	10 mL	√	—	关注		√	√		1	1	1	低	大量冲洗
	$Na_2S_2O_3$	—	0.05 M	3 mL	√	—	关注		√	√		1	1	1	低	大量冲洗
	KNO_3	5.1	0.2 M	7.5 mL	√	眼和皮肤刺激	关注		√	√		1	3	3	低	大量冲洗
	$(NH_4)_2SO_4$	—	0.2 M	7.5 mL	√	—	关注			√		1	1	1	低	大量冲洗
	淀粉溶液	—	0.2%	1 mL	—	—										大量冲洗

B：温度对化学反应速率的影响，求活化能

过程步骤	化合物	类别	浓度 用量	SDS	伤害种类	降低风险	技术	眼镜	手套	通风橱	可能性	危害	风险	级别	措施
按表4-1中实验1的试剂用量分别在高于室温5℃，10℃和15℃的温度下继续进行实验。将测得的各温度下的反应时间，计算得出的各温度下的反应速率及速率常数记录于表4-2。	加热装置				电击&烫伤	确保电线已贴上标签	√	√			1	3	3	低	立刻向指导老师或实验备人员报告

续表

风险评估及控制工作表

实验任务	化学反应速率和活化能				评估人	吴泽颖、王红松、吴星星				评估日期	2021 年 7 月 1 日	
过程步骤	危害鉴别			化学品安全技术说明书	健康风险		安全措施			风险评估		措施
	化合物名称	类别	浓度	用量		可能的伤害种类	降低或消除风险	技术	防护眼镜	手套	通风橱	可能性	危害	风险	危害级别	紧急情况/泼洒 特殊备注
		或物理危害										1~4	1~4		高中低	
C: 催化剂对化学反应速率的影响																
在室温下，按表 4-1 中实验 1 的试剂用量，再分别加入 1 滴、5 滴、10 滴 $Cu(NO_3)_2$ 溶液，不足 10 滴的实验迅速用 $(NH_4)_2S_2O_8$ 溶液补充至 10 滴，搅拌，计时，把此时实验的反应速率与表 4-1 中实验 1 的反应速率比较。	$Cu(NO_3)_2$	—	0.02 M	10 mL	眼和皮肤刺激		关注		√	√	√	1	3	3	低	用大量水清洗
实验产物																
实验废弃物																
整个过程的最终评估	低风险（按照正确流程操作）															
签名	实验老师：				实验室主任/职业安全健康(OHS)代表：							学生/操作人：				

Experiment 4　Determination of chemical reaction rate and activation energy

【Objectives】

1. To determine the effects of concentration, temperature and catalyst on the rate of a chemical reaction.
2. To determine the rate of the reaction between ammonium persulfate and potassium iodide.
3. To calculate the reaction order and the activation energy.

【Principles】

The reaction of ammonium persulfate and potassium iodide is:

$$S_2O_8^{2-} + 3I^- = 2SO_4^{2-} + I_3^- \qquad (4-1)$$

The rate equation for this reaction is:

$$v = k \cdot C^m(S_2O_8^{2-}) \cdot C^n(I^-)$$

Where v is the reaction rate, and k is the rate constant, $C(S_2O_8^{2-})$ and $C(I^-)$ are instant concentrations, the sum of m and n is the reaction order.

The average reaction rate ($\bar{v}$) per unit time can be determined by experiments. If the change of concentration for $S_2O_8^{2-}$ is $\Delta C(S_2O_8^{2-})$ in a period of time (Δt), the average rate can be expressed as:

$$\bar{v} = \Delta C(S_2O_8^{2-})/\Delta t$$

When $\Delta t \to 0$, the average rate can be regarded as the approximate instantaneous rate.

$$\Delta C(S_2O_8^{2-})/\Delta t \approx v = k \cdot C^m(S_2O_8^{2-}) \cdot C^n(I^-)$$

In order to determine the concentration of $S_2O_8^{2-}$, a constant known amount of $Na_2S_2O_3$ and starch solution are added when mixing KI with $(NH_4)_2S_2O_8$. Then two reactions, (4-1) and (4-2), occur at the same time.

$$2S_2O_3^{2-} + I_3^- = S_4O_6^{2-} + 3I^- \qquad (4-2)$$

Reaction (4-2) is very fast and can be completed instantly, while reaction (4-1) is much slower than reaction (4-2). I_3^- produced by reaction (4-1) reacts with $S_2O_3^{2-}$ immediately to form colorless $S_4O_6^{2-}$ and I^-. Therefore, in the beginning of the reaction, the unique blue color of I_3^- and starch could not be seen. Once $Na_2S_2O_3$ is used up, I_3^- still being formed by reaction (4-1) will react with starch and appear blue.

From reaction (4-1) and (4-2), we can find that each mole of $S_2O_3^{2-}$ used is equivalent to 1/2 mole of $S_2O_8^{2-}$ used.

In this experiment, the initial concentration of $Na_2S_2O_3$ is the same for each mixture. Since the blue solution indicates that $S_2O_3^{2-}$ is used up, the initial concentration of $Na_2S_2O_3$ can be used to calculate $\Delta C(S_2O_3^{2-})$ and the average reaction rate.

In logarithmic form, the rate equation becomes:

$$\lg v = m \lg C(S_2O_8^{2-}) + n \lg C(I^-) + \lg k$$

When $C(I^-)$ is kept constant, a plot of $\lg v$ versus $\lg C(S_2O_8^{2-})$ produces a straight line with a slope equal to m. In the same way, a plot of $\lg v$ versus $\lg C(I^-)$ produces a straight line with a slope equal to n when $C(S_2O_8^{2-})$ is kept constant.

Then the value of k, the reaction rate constant, can be calculated with the already known m, n, $C(S_2O_8^{2-})$, $C(I^-)$ and v.

According to the Arrhenius equation,

$$\lg\{k\} = -(E_a/2.303RT) + A$$

where k is rate constant, $kJ \cdot mol^{-1}$; E_a is activation energy ($kJ \cdot mol^{-1}$); R is gas constant ($R = 8.314\ J \cdot K^{-1} \cdot mol^{-1}$); T is thermodynamic temperature; A is constant.

Determine the different k values at different temperatures. A plot of $\lg k$ versus $1/T$ produces a straight line with a slope of $\left(\dfrac{-E_a}{2.303R}\right)$, so the activation energy can be calculated.

【Apparatus and Chemicals】

1. Apparatus

Beakers (100 mL); large test tube; volumetric cylinder; thermostatic water bath; stopwatch; thermometer.

2. Chemicals

$(NH_4)_2S_2O_8$ ($0.2\ mol \cdot L^{-1}$; freshly prepared); KI ($0.2\ mol \cdot L^{-1}$); KNO_3 ($0.2\ mol \cdot L^{-1}$); $Na_2S_2O_3$ ($0.05\ mol \cdot L^{-1}$); $(NH_4)_2SO_4$ ($0.2\ mol \cdot L^{-1}$); $Cu(NO_3)_2$ ($0.02\ mol \cdot L^{-1}$); starch solution (2%, wt/wt).

【Experimental Procedures】

1. Effect of concentration on the reaction rate

At room temperature, the dosage of reagents listed in table 4-1 shall be accurately measured by the volumetric cylinder. Except for $(NH_4)_2S_2O_8$ solution, other reagents can be mixed in the beaker with corresponding number according to the dosage. When $(NH_4)_2S_2O_8$ solution is added, the time shall be counted immediately to mix the solution evenly. When the solution turns blue, the time shall be stopped, and the time Δt and the room temperature shall be recorded.

Table 4-1 Effect of concentration on the reaction rate Room temperature: _____ ℃

	Experiment number	1	2	3	4	5
V/mL	$(NH_4)_2S_2O_8$ (0.2 mol·L^{-1})	10	5	2.5	10	10
	KI (0.2 mol·L^{-1})	10	10	10	5	2.5
	$Na_2S_2O_3$ (0.05 mol·L^{-1})	3	3	3	3	3
	KNO_3 (0.2 mol·L^{-1})				5	7.5
	$(NH_4)_2SO_4$ (0.2 mol·L^{-1})		5	7.5		
	starch solution (0.2%)	1	1	1	1	1
$C_0(S_2O_8^{2-})$/(mol·L^{-1})						
$C_0(I^-)$/(mol·L^{-1})						
$C_0(S_2O_3^{2-})$/(mol·L^{-1})						
Δt/s						
$\Delta(S_2O_3^{2-})$/(mol·L^{-1})						
v/(mol·L^{-1}·s^{-1})						
k/[(mol·L^{-1})$^{1-m-n}$·s^{-1}]						

2. Effect of temperature on the reaction rate

According to the reagent dosage of Experiment 1 in Table 4-1, the experiment was continued at 5 ℃, 10 ℃ and 15 ℃ higher than the room temperature. Record the measured reaction time, calculate reaction rate and rate constant at each temperature in Table 4-2.

Table 4-2 Effect of temperature on the reaction rate

Experiment number	T/K	Δt/s	v/(mol·L^{-1}·s^{-1})	k/[(mol·L^{-1})$^{1-m-n}$·s^{-1}]	$\lg\{k\}$
1					
6					
7					
8					

3. Effect of catalyst on the reaction rate

At room temperature, add 1 drop, 5 drops and 10 drops of $Cu(NO_3)_2$ solution respectively according to the reagent dosage of Experiment 1 in Table 4-1. For the experiment with less than 10 drops, quickly add $(NH_4)_2S_2O_8$ solution to 10 drops, stir, time, and compare the reaction rate of this experiment with that of Experiment 1 in Table 4-1.

Table 4-3 Effect of catalyst on the reaction rate

Experiment number	9	10	11
Drop number of $Cu(NO_3)_2$ solution added			
Δt/s			
v/(mol·L^{-1}·s^{-1})			

【Questions】

1. Why can the reaction rate be calculated by the time lapse from the beginning of the reaction till the deep blue color appears in the experiment? Do the reactions continue or cease when the deep blue color appears?

2. How will it affect the results if the amount of $Na_2S_2O_3$ is too much or too little?

Risk Assessment and Control Worksheet

TASK	Determination of a rate law and activation energy					Assessor	Zeying WU, Hongsong WANG, Xingxing WU					Date	1 JUL. 2021			
Process Steps	Hazard Identification					Health Risk		Safety Measures				Checklist～Risk Assessment				Action
	Chemical Name	Class	Concn.	Amount	MSDS	Type of Injury Possible	To reduce or eliminate the risk	Technique	Glasses	Gloves	Fume Cbd.	Likely 1～4	Hazard 1～4	Risk Score	Risk Level H M L	In case of Emergency/Spill Special Comments
A: Effect of concentration on the reaction rate																
At room temperature, the dosage of reagents listed in table 4-1 shall be accurately measured by the volumetric cylinder. Except for $(NH_4)_2S_2O_8$ solution, other reagents can be mixed in the beaker with corresponding number according to the dosage. When $(NH_4)_2S_2O_8$ solution is added, the time shall be counted immediately to mix the solution evenly. When the solution turns blue, the time shall be stopped, and the time Δt and the room temperature shall be recorded.	$(NH_4)_2S_2O_8$	5.1	0.2 M	10 mL	✓	Skin irritant	care		✓	✓		1	3	3	L	Sponge & water
	KI	—	0.2 M	10 mL	✓	—	care		✓	✓		1	1	1	L	Sponge & water
	$Na_2S_2O_3$	—	0.05 M	3 mL	✓	—	care			✓		1	1	1	L	Sponge & water
	KNO_3	5.1	0.2 M	7.5 mL	✓	Eye & skin irritant	care			✓		1	3	3	L	Sponge & water
	$(NH_4)_2SO_4$	—	0.2 M	7.5 mL	✓	—	care			✓		1	1	1	L	Sponge & water
	starch solution	—	0.2%	1 mL	—	—						1	1	1	L	Sponge & water
B: Effect of temperature on the reaction rate																
According to the reagent dosage of Experiment 1 in Table 4-1, the experiment was continued at 5 ℃, 10 ℃ and 15 ℃ higher than the room temperature. Record the measured reaction time, calculate reaction rate and rate constant at each temperature in Table 4-2.	Heating device					Electric shock & scald	Ensure cord is tagged	✓	✓			1	3	3	L	Report immediately to Demonstrater or Prep Room

Risk Assessment and Control Worksheet

TASK	Determination of a rate law and activation energy					Assessor	Zeying WU, Hongsong WANG, Xingxing WU				Date	1 JUL. 2021	
Process Steps	Hazard Identification					Health Risk		Safety Measures			Checklist～Risk Assessment		Action
	Chemical Name	Class	Concn.	Amount	MSDS	Type of Injury Possible	To reduce or eliminate the risk	Technique	Glasses	Gloves	Fume Cbd.	Risk Assessment	In case of Emergency/Spill
		or Physical Hazard										Likely	Hazard	Risk Score	Risk Level H M L	Special Comments
												1～4	1～4			

Process Steps	Chemical Name	Class	Concn.	Amount	MSDS	Type of Injury Possible	To reduce or eliminate the risk	Technique	Glasses	Gloves	Fume Cbd.	Likely	Hazard	Risk Score	Risk Level	In case of Emergency/Spill	Special Comments
C: Effect of catalyst on the reaction rate																	
At room temperature, add 1 drop, 5 drops and 10 drops of Cu(NO_3)$_2$ solution respectively according to the reagent dosage of Experiment 1 in Table 4-1. For the experiment with less than 10 drops, quickly add (NH_4)$_2$$S_2$$O_8$ solution to 10 drops, stir, time, and compare the reaction rate of this experiment with that of Experiment 1 in Table 4-1.	Cu(NO_3)$_2$	—	0.02 M	10 mL	√	Eye & skin irritant	care		√	√		1	3	3	L	Sponge & water	
Experimental Product (s)																	
Experimental Waste (s)																	
Final assessment of overall process	Low risk when correct procedures followed																
Signatures	Supervisor:					Lab Manager/OHS Represeentative:						Student/Operator:					

实验 5　弱酸电离常数测定（pH 法）

【实验目的】

1. 学习弱酸电离常数测定方法。
2. 学习酸度计的使用方法。
3. 进一步了解电离平衡的基本概念。

【实验原理】

醋酸（简写为 HAc）是一元弱酸，在水溶液中存在下列平衡：

$$HAc \rightleftharpoons H^+ + Ac^-$$

其电离常数表达式为：

$$K_{HAc} = \frac{[H^+][Ac^-]}{[HAc]} \tag{5-1}$$

设醋酸的起始浓度为 c_0，平衡时 $[H^+]=[Ac^-]=x$，代入式（5-1），可以得到：

$$K_{HAc} = \frac{x^2}{c_0 - x} \tag{5-2}$$

在一定温度下，用酸度计测定一系列已知浓度的醋酸的 pH 值，根据 $pH = -\lg[H^+]$，换算出 $[H^+]$，代入式（5-2）中，可求一系列对应的 K 值，取其平均值，即为该温度下醋酸的电离常数。

【器材与试剂】

1. 器材：pHS-3C（或其他型号）酸度计，50 mL 烧杯，酸式滴定管，碱式滴定管，洗耳球等。
2. 试剂：HAc（0.100 0 mol·L^{-1}）标准溶液。

【实验内容】

1. 先将酸度计预热 30 min。
2. 将酸、碱滴定管分别用蒸馏水洗 3 次，碱式滴定管内装蒸馏水至刻度 0.00，酸式滴定管用 HAc 溶液洗 3 次后，装入该 HAc 溶液至刻度 0.00。
3. 不同浓度醋酸溶液的配制：将 5 只洗净烘干的烧杯编成 1～5 号，按表 5-1 配制醋酸溶液。

表 5-1　实验数据记录

编号	HAc 体积(已标定)/mL	H₂O 体积/mL	配制 HAc 浓度/mol·L⁻¹	HAc 的 pH 值
1	3.00	45.00		
2	6.00	42.00		
3	12.00	36.00		
4	24.00	24.00		
5	48.00	0.00		

4. 数据处理。根据实验数据填写表 5-2。

测定时的温度＝_____℃。

表 5-2　实验数据和计算结果

编号	HAc 体积/mL	H₂O 体积/mL	配制[HAc]/mol·L⁻¹	pH 值	[H]/mol·L⁻¹	$K=x^2/(c-x)$
1						
2						
3						
4						
5						

最后计算出 K 的平均值为_____。

【思考题】

1. 本实验测定 HAc 电离常数的依据是什么？
2. pHS-3C 酸度计使用的关键是什么？
3. 怎样配好 HAc 溶液？又如何从 pH 值测得其电离常数？
4. 用 pH 计测定溶液 pH 时，各用什么标准溶液定位？

● 配套课件

风险评估及控制工作表

实验任务	弱酸电离常数测定（pH法）				评估人	吴泽颖，王红松，吴星星				评估日期	2021年7月1日			
过程步骤	危害鉴别				健康风险		安全措施			清单/风险评估	风险评估			措施
	化合物名称	类别	浓度	用量	化学品安全技术说明书	可能的伤害种类	降低或消除风险	技术	防护眼镜	手套	通风橱		可能性 1~4	危害 1~4	风险	危害级别 高中低	紧急情况/泼洒 特殊备注
		或物理危害															
I 将酸度计预热30分钟。	pH酸度计					触电	明确标识使用说明						1	1	1	低	向相关老师报告
II 将酸碱、碱滴定管分别用装蒸馏水至刻度3次，碱式滴定管内装蒸馏水溶液洗3次，酸式滴定管用装HAc溶液洗3次，装入该HAc溶液至刻度0.000。	醋酸	8+3	0.1 M	200 mL	√	皮肤灼伤和眼损伤	关注		√	√			1	1	1	低	用大量水冲洗
实验产物																	
实验废弃物																	
整个过程的最终评估	低风险（按照正确流程操作）																
签名	实验老师：				实验室主任/职业安全健康（OHS）代表：							学生/操作人：					

Experiment 5　Determination of ionization constant of weak acid(pH method)

【Objectives】

1. To learn the method of determining the ionization constant of a weak acid.
2. To learn how to use the pH meter.
3. To further understand the concept of ionization equilibrium.

【Principles】

Acetic acid (HAc) is a weak monacid. The ionization equilibrium of acetic acid is

$$HAc \rightleftharpoons H^+ + Ac^-$$

The ionization constant of acetic acid is expressed as

$$K_{HAc} = \frac{[H^+][Ac^-]}{[HAc]} \tag{5-1}$$

Suppose that the initial concentration of acetic acid is c_0, and the equilibrium concentration of H^+ equals that of Ac^-, that is $[H^+]=[Ac^-]=x$. So, equation (5-1) can be written as

$$K_{HAc} = \frac{x^2}{c_0 - x} \tag{5-2}$$

At a certain temperature, if we prepare the acetic acid solution and determine the pH of this equilibrium solution, we can calculate the concentration of H^+ according to the formula, $pH = -\lg[H^+]$. Then, the ionization constant K (average) can be calculated.

【Apparatus and Chemicals】

1. Apparatus

pH meter (pHS-3C); Geiser burette; Mohr burette; beakers (50 mL); rubber suction bulb and so on.

2. Chemicals

HAc (0.100 0 mol · L^{-1}).

【Experimental Procedures】

1. Preheat the pH meter for 30 minutes.

2. Rinse the two burettes with distilled water three times, and fill the Geiser burette with distilled water. Then, rinse the Mohr burette with acetic acid solution with known concentration, and fill the Mohr burette with acetic acid solution.

3. Prepare acetic acid solutions with different concentrations in five beakers according to Table 5-1.

Table 5-1 Experimental data

Trial	Volume of HAc/mL	Volume of H_2O/mL	Concentration of HAc/mol · L^{-1}	pH value of HAc
1	3.00	45.00		
2	6.00	42.00		
3	12.00	36.00		
4	24.00	24.00		
5	48.00	0.00		

4. Calculate and record data in Table 5-2.

Temperature = _____ ℃.

Table 5-2 Data and calculation

Trial	Volume of HAc/mL	Volume of H_2O/mL	[HAc]/mol · L^{-1}	pH	[H^+]/mol · L^{-1}	$K = x^2/(c-x)$
1						
2						
3						
4						
5						

The average of K is _____.

【Questions】

1. How do you determine the ionization constant of HAc in this experiment?
2. What is the key to using the pH meter?
3. How do you prepare an acetic acid solution? How do you determine its ionization constant from the pH value?
4. What is the standard solution used when determining the pH value with the pH meter?

Risk Assessment and Control Worksheet

TASK	Determination of ionization constant of acetic acid				Assessor	Zeying WU, Hongsong WANG, Xingxing WU				Date	1 JUL. 2021			
Process Steps	Hazard Identification				Health Risk	Safety Measures				Checklist～Risk Assessment			Action	
	Chemical Name	Class or Physical Hazard	Concn.	Amount	MSDS	Type of Injury Possible	To reduce or eliminate the risk	Technique	Glasses	Gloves	Fume Cbd.	Likely	Hazard	Risk Score	Risk Level H M L	In case of Emergency/Spill
															Special Comments
I) Preheat the pH meter for 30 minutes.	pH meter					Electric shock	Ensure cord is tagged					1～4	1～4	1	L	Report immediately to Demonstrator or Prep Room
II) Rinse the two burettes with distilled water three times, and fill the Geiser burette with distilled water. Then, rinse the Mohr burette with acetic acid solution with known concentration, and fill the Mohr burette with acetic acid solution.	Acetic acid	8＋3	0.1 M	200 mL	√	Skin burns and eye injuries	Handle with care		√	√		1	1	1	L	Sponge & water
Experimental Product (s)																
Experimental Waste (s)																
Final assessment of overall process	Low risk when correct procedures followed															
Signatures	Supervisor:					Lab Manager/OHS Representative:						Student/Operator:				

实验6 酸碱反应与缓冲溶液

【实验目的】

1. 理解和巩固酸碱反应的有关概念和原理,包括同离子效应盐类水解及其影响因素等。
2. 学习试管实验的一些基本操作。
3. 了解缓冲溶液的缓冲性能。

【实验原理】

1. 同离子效应

强电解质在水中全部解离。弱电解质在水中部分解离。在一定温度下,弱酸弱碱的解离平衡如下:

$$HA(aq) + H_2O(l) \longleftrightarrow H_3O^+(aq) + A^-(aq)$$

$$B(aq) + H_2O(l) \longleftrightarrow BH^+(aq) + OH^-(aq)$$

在弱电解质溶液中,加入与弱电解质含有相同离子的强电解质,解离平衡向生成弱电解质的方向移动,使弱电解质的解离度下降。这种现象称为同离子效应。

2. 盐类水解

强酸、强碱盐在水中不水解。强酸弱碱盐(如 NH_4Cl)水解溶液显酸性,强碱弱酸盐(如 $NaAc$)水解溶液显碱性。弱酸弱碱盐(如 NH_4Ac)水解溶液的酸碱性取决于弱酸弱碱的相对强弱。例如

$$Ac^-(aq) + H_2O(l) \longleftrightarrow HAc(aq) + OH^-(aq)$$

$$NH_4^+(aq) + Ac^-(aq) + H_2O(l) \longleftrightarrow NH_3 \cdot H_2O(aq) + HAc(aq)$$

水解反应是酸碱中和反应的逆反应。中和反应是放热反应,水解反应是吸热反应。因此升高温度有利于盐类的水解。

3. 缓冲溶液

由弱酸(或弱碱)与弱酸(或弱碱)盐(如 $HAc-NaAc$;$NH_3 \cdot H_2O-NH_4Cl$;$H_3PO_4-NaH_2PO_4$;$NaHPO_4$;$NaHPO_4-Na_3PO_4$ 等)组成的溶液具有保持溶液 pH 相对稳定的性质,这类溶液称为缓冲溶液。

由弱酸弱碱盐组成的缓冲溶液的 pH 可由下列公式计算

$$pH = pK_a^{\ominus}(HA) - \lg c(HA)/c(A^-)$$

由弱酸-弱碱盐组成的缓冲溶液的 pH 可用下式计算:

$$pH = 14 - pK_b^{\ominus}(B) + \lg c(B)/c(BH)$$

缓冲溶液的 pH 可以用 pH 试纸来测定。

缓冲溶液的缓冲能力与组成溶液的弱酸(或弱碱)及其共轭碱(或酸)的浓度有关,当弱碱(或弱酸)与它的共轭碱(或酸)浓度较大时,其缓冲溶液能力较强。此外,缓冲能力还与 $c(HA)/c(A)$ 或 $c(B)/c(BH)$ 有关。当比值接近1时,其缓冲能力最强,此值通常选在 0.1～10 范围内。

【器材与试剂】

1. 器材:pHS - 3C 型(或其他型号)酸度计,10 mL 量筒,50 mL 烧杯,点滴板,试管,试管架,石棉网,酒精灯,pH 试纸。

2. 试剂:HCl 溶液 ($0.1\ mol \cdot L^{-1}$, $2\ mol \cdot L^{-1}$),HAc ($0.1\ mol \cdot L^{-1}$, $1\ mol \cdot L^{-1}$),NaOH ($0.1\ mol \cdot L^{-1}$),$NH_3 \cdot H_2O$ ($0.1\ mol \cdot L^{-1}$, $1\ mol \cdot L^{-1}$),NaCl ($0.1\ mol \cdot L^{-1}$),Na_2CO_3 ($0.1\ mol \cdot L^{-1}$),NH_4Cl ($0.1\ mol \cdot L^{-1}$, $1\ mol \cdot L^{-1}$),NaAc ($1.0\ mol \cdot L^{-1}$),NH_4Ac (s),$BiCl_3$ ($0.1\ mol \cdot L^{-1}$),$CrCl_3$ ($0.1\ mol \cdot L^{-1}$),$Fe(NO_3)_3$ ($0.5\ mol \cdot L^{-1}$),酚酞,甲基橙,未知液 A、B、C、D。

【实验内容】

1. 同离子效应

(1) 用 pH 试纸,酚酞试剂测定和检查 $0.1\ mol \cdot L^{-1}\ NH_3 \cdot H_2O$ 的 pH 及其酸碱性;再加入少量 $NH_4Ac(s)$,观察现象,写出反应的化学方程式,并简要解释。

(2) 用 $0.1\ mol \cdot L^{-1}$ HAc 代替 $0.1\ mol \cdot L^{-1}\ NH_3 \cdot H_2O$,用甲基橙代替酚酞,检测其酸碱度,同样加入少量 $NH_4Ac(s)$ 后观察变化,并简要解释原因。

2. 盐类的水解

(1) A、B、C、D 是四种失去标签的盐溶液,只知它们是 $0.1\ mol \cdot L^{-1}$ 的 NaCl、NaAc、NH_4Cl、Na_2CO_3 溶液,试通过测定其 pH 并结合理论计算确定 A、B、C、D 各为何物。

(2) 观察常温和加热情况下 $0.5\ mol \cdot L^{-1}\ Fe(NO_3)_3$ 的水解情况并解释现象。

(3) 在 3 mL H_2O 中加一滴 $0.1\ mol \cdot L^{-1}\ BiCl_3$ 溶液,观察现象。再缓慢滴加 $2\ mol \cdot L^{-1}$ HCl 溶液,观察有何变化,解释原因。

(4) 在试管中加入 2 滴 $0.1\ mol \cdot L^{-1}\ CrCl_3$ 溶液和 3 滴 $0.1\ mol \cdot L^{-1}\ Na_2CO_3$ 溶液,解释观察到的现象。

3. 缓冲溶液

(1) 按下表中试剂用量配制 4 种缓冲溶液,用酸度计分别测定其 pH,并与计算值进行比较,解释原因。

表 6-1 几种缓冲溶液的 pH

编号	配制缓冲溶液(用量筒量取)	pH 计算值	pH 测定值
1	10.0 mL 1 $mol \cdot L^{-1}$ HAc+10.0 mL 1 $mol \cdot L^{-1}$ NaAc		
2	10.0 mL 0.1 $mol \cdot L^{-1}$ HAc+10.0 mL 1 $mol \cdot L^{-1}$ NaAc		
3	10.0 mL 0.1 $mol \cdot L^{-1}$ HAc 中加入 2 滴酚酞,滴加 0.1 $mol \cdot L^{-1}$ NaOH 溶液至酚酞变红,30 s 不消失,再加入 10.0 mL 0.1 $mol \cdot L^{-1}$ HAc		
4	10.0 mL 1 $mol \cdot L^{-1}$ $NH_3 \cdot H_2O$+10.0 mL 1 $mol \cdot L^{-1}$ NH_4Cl		

(2) 在 1 号缓冲溶液中加入 0.5 mL(约 10 滴)$0.1\ mol \cdot L^{-1}$ HCl 溶液摇匀,用酸度计测定其 pH;再加入 1 mL(约 20 滴)$0.1\ mol \cdot L^{-1}$ NaOH 溶液摇匀,测定其 pH,并与计算值进行比较。

【思考题】

1. 影响盐类水解的因素有哪些?
2. 如何配制 $SbCl_3$ 溶液、$SnCl_2$ 溶液和 $Bi(NO_3)_3$ 溶液？写出它们水解反应的方程式。
3. 缓冲溶液的 pH 由哪些因素决定？其中主要的决定因素是什么？

● 缓冲溶液

风险评估及控制工作表

酸碱反应与缓冲溶液　　　评估人：吴泽颖、王红松、吴星星　　　评估日期：2021年7月1日

实验任务	过程步骤	化合物名称	危害鉴别 类别	危害鉴别 浓度	危害鉴别 用量	化学品安全技术说明书	健康风险 可能的伤害种类	降低或消除风险	安全措施 技术	安全措施 防护眼镜	安全措施 手套	安全措施 通风橱	风险评估清单/风险评估	风险评估 可能性 1~4	风险评估 危害 1~4	风险评估 风险	危害级别 高中低	紧急情况/泼洒 措施	特殊备注
A：同离子效应	Ⅰ）用pH试纸、酚酞试剂测定和检查0.1 mol·L^{-1} NH$_3$·H$_2$O的pH及其酸碱性。	NH$_3$·H$_2$O	8	0.1 M	10 mL	√	皮肤灼伤和眼损伤	关注		√				1	3	3	低	大量水冲洗	
	Ⅱ）用0.1 mol·L^{-1} HAc代替0.1 mol·L^{-1} NH$_3$·H$_2$O，用甲基橙代替酚酞，重复实验Ⅰ）。	HAc	8+3	0.1 M	10 mL	√	皮肤灼伤和眼损伤	关注		√	√			1	1	1	低	大量水冲洗	
B：盐类的水解	Ⅰ）A,B,C,D是四种失去标签的盐溶液，只知它们是0.1 mol·L^{-1}的NaCl、NaAc、NH$_4$Cl、Na$_2$CO$_3$溶液，试通过测定其pH并结合理论计算确定A,B,C,D各为何物。																		
	Ⅱ）试验常温和加热情况下0.5 mol·L^{-1} Fe(NO$_3$)$_3$的水解情况。	酒精灯					烧伤	关注	√					1	3	3	低	立刻向带导老师或准备实验人员报告	
		Fe(NO$_3$)$_3$	3	0.5 M	10 mL	√	皮肤刺激	关注		√	√			1	1	1	低	大量水冲洗	
	Ⅲ）在3 mL H$_2$O中加1滴0.1 mol·L^{-1} BiCl$_3$溶液，观察现象。再滴加2 mol·L^{-1} HCl溶液，观察有何变化。	BiCl$_3$	—	0.1 M	1滴	√	皮肤刺激	关注		√	√			1	1	1	低	大量水冲洗	
		盐酸	2,3+8	2 M	1 mL	√	皮肤灼伤和眼损伤	关注		√	√			1	3	3	低	大量水冲洗	

续 表

风险评估及控制工作表

实验任务	酸碱反应与缓冲溶液					评估人	吴泽颖、王红松、吴星星				评估日期	2021年7月1日				
过程步骤	危害鉴别					健康风险		安全措施			清单/风险评估	风险评估				紧急情况/泄洒 特殊备注
	化合物名称	类别	浓度	用量	化学品安全技术说明书	可能的伤害种类	降低或消除风险	技术	防护眼镜	手套	通风橱		可能性	危害	风险	危害级别
													1～4	1～4		高中低
Ⅳ）在试管中加入 2 滴 0.1 mol·L⁻¹ CrCl₃ 溶液和 3 滴 0.1 mol·L⁻¹ Na₂CO₃ 溶液，观察现象。	氯化铬	—	0.1 M	0.04 mL	√	皮肤过敏	关注		√	√			1	1	1	低 大量水冲洗
	Na₂CO₃溶液	—	0.1 M	3 滴	√	眼刺激	关注		√	√			1	1	1	低 大量水冲洗
C: 缓冲溶液																
Ⅰ）按表 6-1 中试剂用量配制 4 种缓冲溶液，用酸度计分别测定其 pH，并与计算值进行比较。																
Ⅱ）10.0 mL 0.1 mol·L⁻¹ HAc 中加入 2 滴酚酞，滴加 0.1 mol·L⁻¹ NaOH 溶液至酚酞变红，30 s 不消失，再加入 10.0 mL 0.1 mol·L⁻¹ HAc。	氢氧化钠	8	0.1 M	5 mL	√	皮肤灼伤和眼损伤	关注		√	√			1	3	3	低 大量水冲洗
实验产物																
实验废弃物																
整个过程的最终评估	低风险（按照正确流程操作）															
签名	实验老师：					实验室主任/职业安全健康（OHS）代表：					学生/操作人：					

Experiment 6 Acid-base reaction and buffer solution

【Objectives】

1. To understand and consolidate the concepts and principles related to acid-base reaction including ion effect and salt hydrolysis.

2. To learn the preparation of buffer solution and the determination of pH, and the use of acidimeter.

3. To understand the buffer performance and principle of buffer solution.

【Principles】

1. Same-ion effect

Strong electrolytes are all dissociated in water. Weak electrolytes are partially dissociated in water. At a certain temperature, the dissociation equilibrium of weak acids and bases is as follows:

$$HA(aq) + H_2O(l) \longleftrightarrow H_3O^+(aq) + A^-(aq)$$

$$B(aq) + H_2O(l) \longleftrightarrow BH^+(aq) + OH^-(aq)$$

In the weak electrolyte solution, when the same ions from a strong electrolyte is added into the weak electrolyte solution, the dissociation equilibrium will move in the direction of generating the weak electrolyte, so that the degree of dissociation of the weak electrolyte decreases. This phenomenon is called the same-ion effect.

2. Salt hydrolysis

Strong acid and alkali salts do not hydrolyze in water. The hydrolysis solution of strong acid weak base salt (such as NH_4Cl) is acidic, and the hydrolysis solution of strong base weak acid salt (such as NaAc) is alkaline. The acidity and alkalinity of a weak acid weak base salt (such as NH_4Ac) hydrolysis solution depends on the relative strength of the weak acid and weak base. E.g

$$Ac^-(aq) + H_2O(l) \longleftrightarrow HAc(aq) + OH^-(aq)$$

$$NH_4^+(aq) + Ac^-(aq) + H_2O(l) \longleftrightarrow NH_3 \cdot H_2O(aq) + HAc(aq)$$

The hydrolysis reaction is the reverse reaction of the acid-base neutralization reaction. The neutralization reaction is an exothermic reaction, and the hydrolysis reaction is an endothermic reaction. Therefore increasing the temperature is favorable for the hydrolysis of the salts.

3. Buffer solution

A solution consisting of a weak acid (or a weak base) and a weak acid (or a weak base) salt (such as $HAc-NaAc$; $NH_3 \cdot H_2O - NH_4Cl$; $H_3PO_4 - NaH_2PO_4$; $NaHPO_4$; $NaHPO_4 - Na_3PO_4$ etc.) has the property of keeping the pH of the solution relatively stable. This type of solution is called a buffer solution.

The pH of the buffer solution consisting of weak acid and base salts can be calculated by the following formula

$$pH = pK_a^{\ominus}(HA) - \lg c(HA)/c(A^-)$$

$$pH = 14 - pK_b^{\ominus}(B) + \lg c(B)/c(BH)$$

The pH of the buffer solution can be measured using a pH test paper.

The buffering capacity of a buffer solution is related to the concentration of the weak acid (or weak base) and its conjugate base (or acid) that make up the solution. When the concentration of the weak base (or weak acid) and its conjugate base (or acid) is large, its buffer solution capacity is strong. In addition, the buffering capacity is also related to $c(HA)/c(A)$ 或 $c(B)/c(BH)$. When the ratio is close to 1, its buffering capacity is the strongest. This value is usually selected in the range of 0.1~10.

【Apparatus and Chemicals】

1. Apparatus

pHS-3C (or other models) acidity meter; 5 simple (10 mL); 4 beaker (50 mL); drip plate; test tube; test tube rack; asbestos net; alcohol lamp; pH test paper.

2. Chemicals

HCl solution (0.1 mol·L^{-1}, 2 mol·L^{-1}); HAc (0.1 mol·L^{-1}, 1 mol·L^{-1}); NaOH (0.1 mol·L^{-1}); $NH_3 \cdot H_2O$ (0.1 mol·L^{-1}, 1 mol·L^{-1}); NaCl (0.1 mol·L^{-1}); Na_2CO_3 (0.1 mol·L^{-1}); NH_4Cl (0.1 mol·L^{-1}, 1 mol·L^{-1}); NaAc (1.0 mol·L^{-1}); NH_4Ac(s); $BiCl_3$ (0.1 mol·L^{-1}); $CrCl_3$ (0.1 mol·L^{-1}); $Fe(NO_3)_3$ (0.5 mol·L^{-1}); Phenolphthalein; methyl orange; unknown solutions A, B, C, D.

【Experimental Procedure】

1. Same-ion effect

(1) Use pH test paper and phenolphthalein reagent to measure and check the pH of 0.1 mol·L^{-1} $NH_3 \cdot H_2O$; add a small amount of NH_4Ac(s) and observe the phenomenon, write down the reaction equation and explain it briefly.

(2) Replace 0.1 mol·L^{-1} $NH_3 \cdot H_2O$ with 0.1 mol·L^{-1} HAc and phenolphthalein with methyl orange, and repeat operation (1).

2. Hydrolysis of salts

(1) A, B, C, and D are four kinds of salt solutions without labels. You only know that they are 0.1 mol·L^{-1} of NaCl, NaAc, NH_4Cl and Na_2CO_3 solutions. Determine A, B, C and D by measuring its pH and combining theoretical calculation.

(2) Observe the hydrolysis of 0.5 mol·L^{-1} $Fe(NO_3)_3$ under normal temperature and heating

conditions.

(3) Add a drop of 0.1 mol·L^{-1} BiCl$_3$ solution in 3 mL H$_2$O and observe the phenomenon, and then add 2 mol·L^{-1} HCl solution dropwise, observe any changes.

(4) Add 2 drops of 0.1 mol·L^{-1} CrCl$_3$ solution and 3 drops of 0.1 mol·L^{-1} Na$_2$CO$_3$ solution into the test tube, observe the phenomenon.

3. Buffer solution

(1) Prepare 4 kinds of buffer solutions according to the dosage of reagents in the following table, measure their pH with acidity meter, and compare with the calculated value.

(2) Add 0.5 mL (about 10 drops) of 0.1 mol·L^{-1} HCl solution to No. 1 buffer solution and measure its pH with a pH meter, and then add 1 mL (about 20 drops) of 0.1 mol·L^{-1} NaOH into the solution. The pH should be measured and compared with the calculated value.

Table1 6-1 pH of several buffer solutions

num	Blending the buffer solution (measured with a measuring cylinder)	Calculation of pH value	Determination of pH value
1	10.0 mL 1 mol·L^{-1} HAc + 10.0 mL 1 mol·L^{-1} NaAc		
2	10.0 mL 0.1 mol·L^{-1} HAc + 10.0 mL 1 mol·L^{-1} NaAc		
3	Add 2 drops of phenolphthalein to 10.0 mL 0.1 mol·L^{-1} HAc, and then add 0.1 mol·L^{-1} NaOH solution dropwise until the solution turns into red and not disappear in 30 s, then add 10.0 mL 0.1 mol·L^{-1} HAc		
4	10.0 mL 1 mol·L^{-1} NH$_3$H$_2$O + 10.0 mL 1 mol·L^{-1} NH$_4$Cl		

【Questions】

1. What are the factors that affect the hydrolysis of salts?

2. How to prepare SbCl$_3$ solution, SnCl$_2$ solution and Bi(NO$_3$)$_3$ solution? Write down the equation of their hydrolysis reaction.

3. What factors determine the pH of the buffer solution? What are the main determinants?

Risk Assessment and Control Worksheet

TASK	Assessor		Date
Acid-base reaction and Buffer solution	Zeying WU, Hongsong WANG, Xingxing WU		1 JUL. 2021

Process Steps	Hazard Identification				Health Risk		Safety Measures				Checklist — Risk Assessment				Action	
	Chemical Name	Class	Concn.	Amount	MSDS	Type of Injury Possible	To reduce or eliminate the risk	Technique	Glasses	Gloves	Fume Cbd.	Likely 1~4	Hazard 1~4	Risk Score	Risk Level H M L	In case of Emergency/Spill	Special Comments
		or Physical Hazard															

A: Same-ion effect

Process Steps	Chemical	Class	Concn.	Amount	MSDS	Injury	Reduce risk	Tech	Glasses	Gloves	Fume	Likely	Hazard	Score	Level	Action
Ⅰ) Use pH test paper and phenolphthalein reagent to measure and check the pH of 0.1 mol·L⁻¹ NH₃·H₂O	Ammonium hydroxide	8	0.1 M	10 mL	√	Skin burns and eye injuries	Handle with care		√	√	√	1	3	3	L	Sponge & water
Ⅱ) Replace 0.1 mol·L⁻¹ NH₃·H₂O with 0.1 mol·L⁻¹ HAc and phenolphthalein with methyl orange, and repeat operation Ⅰ).	HAc	8+3	0.1 M	10 mL	√	Skin burns and eye injuries	Handle with care		√	√	√	1	1	1	L	Sponge & water

B: Hydrolysis of salts

Process Steps	Chemical	Class	Concn.	Amount	MSDS	Injury	Reduce risk	Tech	Glasses	Gloves	Fume	Likely	Hazard	Score	Level	Action
Ⅰ) A, B, C, and D are four kinds of salt solutions without labels. You only know that they are 0.1 mol·L⁻¹ of NaCl, NaAc, NH₄Cl and Na₂CO₃ solutions. Determine A, B, C and D by measuring its pH and combining theoretical calculation.																
Ⅱ) Observe the hydrolysis of 0.5 mol·L⁻¹ Fe(NO₃)₃ under normal temperature and heating conditions.	Alcohol burner					Burns	Handle with care	√	√	√		1	3	3	L	Report immediately to Demorstrator or Prep Room
	Fe(HO₃)₃	3	0.5 M	10 mL	√	Skin irritant	Handle with care		√	√		1	1	1	L	Sponge & water
Ⅲ) Add a drop of 0.1 mol·L⁻¹ BiCl₃ solution in 3 mL H₂O and observe the phenomenon, and then add 2 mol·L⁻¹ HCl solution dropwise, observe any changes.	BiCl₃ solution	—	0.1 M	1 drop	√	Skin irritant	Handle with care		√	√	√	1	1	1	L	Sponge & water
	HCl	2,3+8	2M	1 mL	√	Skin burns and eye injuries	Handle with care		√	√	√	1	3	3	L	Sponge & water

Risk Assessment and Control Worksheet — continued

TASK	Acid-base reaction and Buffer solution					Assessor	Zeying WU, Hongsong WANG, Xingxing WU				Date	1 JUL. 2021		
Process Steps	Hazard Identification					Health Risk	Safety Measures				Checklist — Risk Assessment			Action
	Chemical Name	Class	Concn.	Amount	MSDS	Type of Injury Possible	To reduce or eliminate the risk	Technique	Glasses	Gloves	Fume Cbd.	Likely	Hazard	Risk Score	Risk Level H M L	In case of Emergency/Spill
		or Physical Hazard													Special Comments
Ⅳ) Add 2 drops of 0.1 M CrCl₃ solution and 3 drops of 0.1 M Na₂CO₃ solution into the test tube, observe the phenomenon.	Chromium trichloride	—	0.1 M	0.04 mL	√	skin allergy	Handle with care		√	√		1~4	1~4	1	L	Sponge & water
	Na₂CO₃ solution	—	0.1 M	3 drops	√	skin irritant	Handle with care		√	√		1	1	1	L	Sponge & water
C: Buffer solution																
Ⅰ) Prepare 4 kinds of buffer solutions according to the dosage of reagents in the table 6 – 1, measure their pH with acidity meter, and compare with the calculated value.																
Ⅱ) Add 2 drops of phenolphthalein to 10.0 mL 0.1 M HAc, and then add 0.1 M NaOH solution dropwise until the solution turns into red and not disappear in 30 s, then add 10.0 mL 0.1 M HAc.	Sodium hydroxide	8	0.1 M	5 mL	√	Skin burns and eye injuries	Handle with care		√	√		1	3	3	L	Sponge & water
Experimental Product (s)																
Experimental Waste (s)																
Final assessment of overall process	Low risk when correct procedures followed															
Signatures	Supervisor:						Lab Manager/OHS Representative:					Student/Operator:				

第 4 章
元素化学实验

实验 7　硼碳硅氮磷

【实验目的】

1. 掌握硼酸和硼砂的重要性质,了解可溶性硅酸盐的水解性和难溶的硅酸盐的生成与颜色。
2. 掌握硝酸、亚硝酸及其盐的重要性质。
3. 了解磷酸盐的主要性质。
4. 学习并掌握 CO_3^{2-}、NH_4^+、NO_2^-、NO_3^-、PO_4^{3-} 的鉴定方法。

【实验原理】

硼酸是一元弱酸,它在水溶液中的解离不同于一般的一元弱酸。硼酸是 lewis 酸,能与多羟基醇发生加合反应,使溶液的酸性增强。

硼砂的水溶液因水解而呈碱性。硼砂溶液与酸反应可析出硼酸。

鉴定 CO_3^{2-} 的常用方法是将碳酸盐溶液与盐酸反应生成的 CO_2 通入 $Ba(OH)_2$ 溶液中,能使 $Ba(OH)_2$ 溶液变浑浊。

大多数硅酸盐难溶于水,硅酸钠可发生水解作用。

鉴定 NH_4^+ 的常用方法有两种,一是 NH_4^+ 与 OH^- 反应,生成的 $NH_3(g)$ 使红色石蕊试纸变蓝;二是 NH_4^+ 与奈斯勒(Nessler)试剂($K_2[HgI_4]$的碱性溶液)反应,生成红棕色沉淀。

亚硝酸盐溶液与强酸反应生成亚硝酸,进而分解为 N_2O_3 和 H_2O。N_2O_3 又能分解为 NO 和 NO_2。亚硝酸盐中氮的氧化值为+3,它在酸性溶液中作氧化剂,可被还原为 NO;与强氧化剂作用时则生成硝酸盐。

硝酸具有强氧化性,可与许多非金属发生氧化还原反应,主要还原产物是 NO。浓硝酸与金属反应主要生成 NO_2,稀硝酸与金属反应通常生成 NO,活泼金属能将稀硝酸还原为 NH_4^+。

NO_2^- 的鉴定方法:NO_2^- 与 $FeSO_4$ 溶液在 HAc 介质中反应生成棕色的 $[Fe(NO)(H_2O)_5]^{2+}$(简写为 $[Fe(NO)]^{2+}$)。

$$Fe^{2+} + NO_2^- + 2HAc \longrightarrow Fe^{3+} + NO + H_2O + 2Ac^-$$

$$Fe^{2+} + NO \longrightarrow [Fe(NO)]^{2+}$$

NO_3^- 的鉴定方法：NO_3^- 与 $FeSO_4$ 溶液在 H_2SO_4 介质中反应生成棕色的 $[Fe(NO)]^{2+}$。

$$3Fe^{2+} + NO_3^- + 4H^+ \longrightarrow 3Fe^{3+} + NO + 2H_2O$$

$$Fe^{2+} + NO \longrightarrow [Fe(NO)]^{2+}$$

在试液与浓硫酸液层界面处生成的 $[Fe(NO)]^{2+}$ 呈棕色环状。此方法称为"棕色环"法。NO_2^- 的存在干扰 NO_3^- 的鉴定，可加入尿素并微热除去 NO_2^-。

$$NO_2^- + CO(NH_2)_2 + 2H^+ \longrightarrow 2N_2 + CO_2 + 3H_2O$$

碱金属（锂除外）和铵的磷酸盐、磷酸一氢盐易溶于水，其他磷酸盐难溶于水。大多数磷酸二氢盐易溶于水。焦磷酸盐和三聚磷酸盐都具有配位作用。

PO_4^{3-} 的鉴定方法：PO_4^{3-} 与 $(NH_4)_2MoO_4$ 溶液在硝酸介质中反应，生成黄色的磷钼酸铵沉淀。

【器材与试剂】

1. 器材：点滴板，水浴锅，pH 试纸，红色石蕊试纸，镍铬丝（一端做成环状）。

2. 试剂：HCl 溶液（2 mol·L^{-1}，6 mol·L^{-1}，浓），H$_2$SO$_4$（1 mol·L^{-1}，6 mol·L^{-1}，浓），HNO$_3$（2 mol·L^{-1}，浓），HAc（2 mol·L^{-1}），NaOH（2 mol·L^{-1}，6 mol·L^{-1}），Ba(OH)$_2$（饱和），Na$_2$CO$_3$（0.1 mol·L^{-1}），NaHCO$_3$（0.1 mol·L^{-1}），Na$_2$SiO$_3$（0.5 mol·L^{-1}，20%），NH$_4$Cl（0.1 mol·L^{-1}），BaCl$_2$（0.5 mol·L^{-1}），NaNO$_2$（0.1 mol·L^{-1}，1 mol·L^{-1}），KI（0.02 mol·L^{-1}），KMnO$_4$（0.01 mol·L^{-1}），KNO$_3$（0.1 mol·L^{-1}），Na$_3$PO$_4$（0.1 mol·L^{-1}），Na$_2$HPO$_4$（0.1 mol·L^{-1}），NaH$_2$PO$_4$（0.1 mol·L^{-1}），CaCl$_2$（0.1 mol·L^{-1}），CuSO$_4$（0.1 mol·L^{-1}），Na$_4$P$_2$O$_7$（0.5 mol·L^{-1}），Na$_5$P$_3$O$_{10}$（0.1 mol·L^{-1}），Na$_2$B$_4$O$_7$·10H$_2$O(s)，H$_3$BO$_3$(s)，Co(NO$_3$)$_2$·6H$_2$O(s)，CaCl$_2$(s)，锌粉，铜屑，CuSO$_4$·5H$_2$O(s)，ZnSO$_4$·7H$_2$O(s)，Fe$_2$(SO$_4$)$_3$(s)，NiSO$_4$·7H$_2$O(s)，FeSO$_4$·7H$_2$O(s)，Co(NH$_2$)$_2$(s)，NH$_4$NO$_3$(s)，Na$_3$PO$_4$·12H$_2$O(s)，NaHCO$_3$(s)，Na$_2$CO$_3$(s)，甘油，甲基橙指示剂，Nessler 试剂，淀粉试液，钼酸铵试剂。

【实验内容】

1. 硼砂和硼酸的性质

（1）取约 0.5 g 的硼酸晶体和 3 mL 的去离子水于试管中，观察溶解情况。微热使其全部溶解，冷却至室温，用 pH 试纸测定溶液的 pH。然后在溶液中加入 1 滴甲基橙指示剂混合均匀。将溶液分成两份，在一份中加入 10 滴甘油，摇匀后比较两份溶液的颜色。

（2）取约 1 g 硼砂和 2 mL 去离子水于试管中，微热使其溶解，用 pH 试纸测定溶液的 pH。加入 1 mL 6 mol·L^{-1} H$_2$SO$_4$ 溶液，用玻璃棒不断搅拌的同时将试管放在冷水中冷却，观察硼酸晶体的析出。

2. CO_3^{2-} 的鉴定

在试管中加入 1 mL 0.1 mol·L^{-1} Na$_2$CO$_3$ 溶液和 1 mL 2 mol·L^{-1} 的 HCl 溶液，用带导管的塞子盖紧试管口，将产生的气体通入 Ba(OH)$_2$ 饱和溶液中，观察现象。

3. 硅酸盐的性质

在试管中加入 1 mL 0.5 mol·L^{-1} 的 Na$_2$SiO$_3$ 溶液，用 pH 试纸测其 pH。然后逐滴加入 6 mol·L^{-1} HCl

溶液,使溶液的 pH 在 6~9,观察硅酸凝胶的生成(若无凝胶生成可微热)。

4. NH_4^+ 的检验

(1) 取少量 0.1 mol·L^{-1} 的 NH_4Cl 溶液和 2 mol·L^{-1} 的 NaOH 溶液于试管中,微热后用湿润的红色石蕊试纸在试管口检验逸出的气体。解释观察到的现象。

(2) 在滤纸条上加 1 滴奈斯勒试剂,代替红色的石蕊试纸重复实验(1),观察现象。

5. 硝酸的氧化性

(1) 在试管内放入 1 小块铜屑,加入几滴浓硝酸,观察现象。

(2) 在试管中放入少量锌粉,加入 1 mL 2 mol·L^{-1} HNO_3 观察现象(如不反应可微热)。

6. NO_3^- 和 NO_2^- 的鉴定

(1) 取 1 mL 0.1 mol·L^{-1} KNO_3 溶液和少量 $FeSO_4·7H_2O$ 晶体于试管中,摇荡使其溶解。斜持试管,沿试管壁小心滴加 1 mL 浓硫酸,静置片刻,观察两种液体界面处的棕色环。

(2) 取 1 滴 0.1 mol·L^{-1} $NaNO_2$ 溶液稀释至 1 mL,加少量 $FeSO_4·7H_2O$ 晶体摇荡试管使其溶解,加入 2 mol·L^{-1} HAc 溶液,观察现象。

(3) 取 0.1 mol·L^{-1} KNO_3 溶液和 0.1 mol·L^{-1} $NaNO_2$ 溶液各 2 滴稀释至 1 mL,再加少量尿素及 2 滴 1 mol·L^{-1} 的 H_2SO_4 以消除 NO_2^- 对鉴定 NO_3^- 的干扰,再进行棕色环试验。

7. 磷酸盐的性质

(1) 用 pH 试纸分别测定 Na_3PO_4、Na_2HPO_4 和 NaH_2PO_4 各 0.1 mol·L^{-1} 溶液的 pH。

(2) 在 3 支试管中分别滴加浓度为 0.1 mol·L^{-1} 的 Na_3PO_4、Na_2HPO_4 和 NaH_2PO_4 溶液 1 mL,再加入几滴 0.1 mol·L^{-1} $CaCl_2$ 溶液,观察现象。

(3) 在试管中加几滴 0.1 mol·L^{-1} $CuSO_4$ 溶液,然后逐滴加入 0.5 mol·L^{-1} Na_4PO_7 溶液至过量,观察现象。

(4) 在试管中加入 1 滴 0.1 mol·L^{-1} $CaCl_2$ 溶液,然后滴加 0.1 mol·L^{-1} Na_2CO_3 溶液,再滴加 0.1 mol·L^{-1} $Na_5P_3O_{10}$ 溶液,观察各自现象。

8. PO_4^{3-} 的测定

取几滴 0.1 mol·L^{-1} $NaPO_3$ 溶液,加入少量 0.5 mol·L^{-1} HNO_3,再加入 1 mL 钼酸铵试剂,在 40~45 ℃ 水浴微热,观察现象。

【思考题】

1. 为什么说硼酸有别于一般的一元弱酸,其酸性从何而来?
2. 为什么在 Na_2SiO_3 溶液中加入 HAc 溶液、NH_4Cl 溶液或通入 CO_2,都能生成硅酸凝胶?
3. 鉴定 NH_4^+ 时,为什么将奈斯勒试剂滴在滤纸上检验逸出的 NH_3 而不是将奈斯勒试剂直接加到含 NH_4^+ 的溶液中?
4. 硝酸与金属反应的主要还原产物与哪些因素有关?
5. NO_3^- 的存在是否干扰 NO_2^- 的鉴定?
6. 用铝酸铵试剂鉴定 PO_4^{3-} 时为什么要在硝酸介质中进行?

- 元素毒性
- 化学元素

风险评估及控制工作表

硼碳硅氮磷　　吴泽颖，王红松，吴星星　　2021年7月1日

实验任务	过程步骤	危害鉴别				评估风险		安全措施			评估清单/风险评估				紧急情况、泼洒措施	特殊备注
		化合物名称	类别	液度	用量	化学品安全技术说明书	或物理危害	可能的伤害种类	降低或消除风险	技术	防护眼镜	手套	通风橱	可能性 1~4	危害 1~4	风险	危害级别 高中低

A: 硼砂和硼酸的性质

I) 在试管中加入约 0.5 g 的硼酸晶体和 3 mL 的去离子水，观察溶解情况。微热至温，用 pH 试纸测定溶解的 pH，冷至室温。然后在溶液中加入 1 滴甲基橙指示剂，并将溶液分成两份，在一份中加入 10 滴甘油，混合均匀，比较两份溶液的颜色。

| 硼酸晶体 | — | — | 0.5 g | √ | | — | 关注 | | √ | √ | | 1 | 1 | 1 | 低 | 大量水冲洗 |
| 加热装置 | | | | | | | 电击 & 烫伤 | 确保电线已贴上标签 | √ | √ | √ | | 1 | 3 | 3 | 低 | 向指导老师或实验准备人员报告 |

II) 在试管中加入约 1 g 硼砂和 2 mL 去离子水，微热使其溶解，用 pH 试纸测定溶液的 pH。然后加入 1 mL 6 mol·L⁻¹ H₂SO₄ 溶液，将试管放在冷水中冷却，并用玻璃棒不断搅拌，片刻后观察硼酸晶体的析出。

| 硫酸 | 8 | 6 M | 1 mL | √ | | 皮肤灼伤和眼其损伤 | 关注 | | √ | √ | √ | 1 | 1 | 1 | 低 | 大量水冲洗 |

B: 碳酸根的鉴定

在试管中加入 1 mL 0.1 mol·L⁻¹ Na₂CO₃ 溶液，再加入半滴管 2 mol·L⁻¹ 的 HCl 溶液，立即用带导气管的塞子盖紧试管口，将产生的气体通入 Ba(OH)₂ 饱和溶液中，观察现象。

Na₂CO₃	—	0.1 M	1 mL	√		眼刺激	关注		√	√	√	1	1	1	低	大量水冲洗
盐酸	2.3+8	2 M	3 mL	√		皮肤灼伤和眼损伤	关注		√	√	√	1	3	3	低	大量水冲洗
Ba(OH)₂ 饱和溶液	8		20 mL	√		皮肤灼伤和眼损伤	关注		√	√	√	1	3	3	低	大量水冲洗

风险评估及控制工作表

吴泽颖，王红松，吴星星　　　　评估日期 2021年7月1日

<table>
<tr><th rowspan="3">实验任务
过程步骤</th><th colspan="4">硼碳硅氮磷
危害鉴别</th><th colspan="3">评估人
健康风险</th><th colspan="4">安全措施</th><th colspan="4">风险评估</th><th rowspan="3">紧急情况/泼洒
措施
特殊备注</th></tr>
<tr><th>化合物名称</th><th>类别</th><th>浓度</th><th>用量</th><th>化学品安全技术说明书</th><th>可能的伤害种类</th><th>降低或消除风险</th><th>技术</th><th>防护眼镜</th><th>手套</th><th>通风橱</th><th>评估清单/风险评估</th><th>可能性</th><th>危害</th><th>风险</th><th>危害级别
高中低</th></tr>
<tr><th colspan="4">或物理危害</th><th></th><th></th><th></th><th></th><th></th><th></th><th></th><th></th><th colspan="3">1～4　1～4</th><th></th></tr>

<tr><td colspan="17">C：硅酸盐的性质</td></tr>
<tr><td>在试管中加入 1 mL 0.5 mol·L⁻¹ 的 Na₂SiO₃ 溶液，用 pH 试纸测其pH。然后逐滴加入 6 mol·L⁻¹ HCl 溶液，使溶液的 pH 在 6～9，观察硅酸凝胶的生成。</td><td>Na₂SiO₃ 溶液</td><td>—</td><td>0.5 M</td><td>1 mL</td><td>√</td><td>—</td><td>关注</td><td></td><td>√</td><td></td><td></td><td></td><td>1</td><td>1</td><td>1</td><td>低</td><td>大量冲洗</td></tr>
<tr><td></td><td>盐酸</td><td>2,3+8</td><td>6 M</td><td>1 mL</td><td>√</td><td>皮肤灼伤和眼损伤</td><td>关注</td><td></td><td>√</td><td>√</td><td></td><td></td><td>1</td><td>3</td><td>3</td><td>低</td><td>大量冲洗</td></tr>

<tr><td colspan="17">D：氨根的检验</td></tr>
<tr><td>I) 在试管中加入少量 0.1 mol·L⁻¹ 的 NH₄Cl 溶液和 2 mol·L⁻¹ 的 NaOH 溶液，微热用湿润的红色石蕊试纸在试管口检验逸出的气体，写出有关反应的化学方程式。</td><td>NH₄Cl 溶液</td><td>—</td><td>0.1 M</td><td>少量</td><td>√</td><td>眼刺激</td><td>关注</td><td></td><td>√</td><td>√</td><td>√</td><td></td><td>1</td><td>2</td><td>2</td><td>低</td><td>大量冲洗</td></tr>
<tr><td></td><td>NaOH 溶液</td><td>8</td><td>2 M</td><td>少量</td><td>√</td><td>皮肤灼伤和眼损伤</td><td>关注</td><td></td><td>√</td><td>√</td><td>√</td><td></td><td>1</td><td>3</td><td>3</td><td>低</td><td>大量冲洗</td></tr>
<tr><td>II) 在滤条上加 1 滴奈斯勒试剂，代替红色的石蕊试纸重复实验(1)，观察现象。</td><td></td><td></td><td></td><td></td><td></td><td></td><td></td><td></td><td></td><td></td><td></td><td></td><td></td><td></td><td></td><td></td><td></td></tr>

<tr><td colspan="17">E：硝酸的氧化性</td></tr>
<tr><td>I) 在试管内放入 1 小块铜屑，加入几滴浓硝酸，观察现象。然后迅速加水稀释，倒掉溶液，回收铜屑。</td><td>铜屑</td><td>—</td><td>—</td><td>少量</td><td>√</td><td>—</td><td></td><td></td><td>√</td><td>√</td><td>√</td><td></td><td>1</td><td>1</td><td>1</td><td>低</td><td>大量冲洗</td></tr>
<tr><td></td><td>硝酸</td><td>8+5.1</td><td>8 M</td><td>1 mL</td><td>√</td><td>皮肤灼伤和眼损伤</td><td>关注</td><td></td><td>√</td><td>√</td><td>√</td><td></td><td>1</td><td>3</td><td>3</td><td>低</td><td>大量冲洗</td></tr>
</table>

续表

风险评估及控制工作表

评估人：吴泽颖、王红松、吴星星 评估日期：2021年7月1日

实验任务	过程步骤	化合物名称	危害鉴别 类别	危害鉴别 浓度	危害鉴别 用量	化学品安全技术说明书	健康风险 可能的伤害种类	降低或消除风险	安全措施 技术	防护眼镜	手套	通风橱	清单风险评估	可能性	危害风险	风险	危害级别	措施
			或物理危害												1~4	1~4	高中低	紧急情况/特殊备注
	Ⅱ）在试管中放入少量锌粉，加入 1 mL 2 mol·L⁻¹ HNO₃观察现象（如不反应可微热），取清液检验是否有 NH₄⁺ 生成	锌粉	4.2+4.3		少量	√	—							1	1	1	低	大量冲洗
		HNO₃	8+5.1	2 M	1 mL	√	皮肤灼伤和眼损伤	关注		√	√	√		1	3	3	低	大量冲洗
F：NO₃⁻和NO₂⁻的鉴定																		
	Ⅰ）取 1 mL 0.1 mol·L⁻¹ KNO₃溶液至 1 mL，加少量 FeSO₄·7H₂O 晶体，摇荡试管使其溶解，然后沿试管壁小心滴加 1 mL 浓硫酸，静置片刻，观察两种液体界面处的棕色环。	KNO₃	5.1	0.1 M	1 mL	√	皮肤刺激	关注		√	√	√		1	1	1	低	大量冲洗
		FeSO₄·7H₂O 晶体	—		少许	√	—	关注		√	√	√		1	1	1	低	大量冲洗
		硫酸	8	12 M	1 mL	√	皮肤灼伤和眼损伤	关注		√	√	√		1	4	4	低	大量冲洗
	Ⅱ）取1滴 0.1 mol·L⁻¹ NaNO₂溶液稀释至 1 mL，加少量 FeSO₄·7H₂O 晶体，加入 2 mol·L⁻¹ HAc 溶液，观察现象。	HAc	8+3	2 M	少许	√	皮肤灼伤和眼损伤	关注		√	√	√		1	1	1	低	大量冲洗
	Ⅲ）取 0.1 mol·L⁻¹ KNO₃溶液和 0.1 mol·L⁻¹ NaNO₂溶液各 2 滴稀释至 1 mL，再加少量尿素及 2 滴 1 mol·L⁻¹ 的 H₂SO₄以消除 NO₂⁻对鉴定 NO₃⁻的干扰，然后进行棕色环试验。	尿素	—		少许	√	—			√	√	√		1	1	1	低	大量冲洗

续表

风险评估及控制工作表 — 棚碳硅氮磷 — 评估人：吴泽颖、王红松、吴星星 — 评估日期：2021年7月1日

实验任务														
过程步骤	化合物名称	危害鉴别 类别	浓度	用量	化学品安全技术说明书	健康风险 可能的伤害种类	或物理危害	降低或消除风险	安全措施 技术	防护眼镜	手套	通风橱	清单/风险评估 可能性 1~4	危害 1~4	风险	危害级别 高中低	紧急情况/泼洒	特殊备注

G：磷酸盐的性质

过程步骤	化合物名称	类别	浓度/用量	化学品安全技术说明书	可能的伤害种类	或物理危害	降低或消除风险	技术	防护眼镜	手套	通风橱	可能性	危害	风险	危害级别	紧急情况/泼洒	特殊备注
Ⅰ）用pH试纸分别测定 0.1 mol·L⁻¹ Na₃PO₄、0.1 mol·L⁻¹ Na₂HPO₄ 和 0.1 mol·L⁻¹ NaH₂PO₄ 溶液的pH。	Na₃PO₄	—	0.1 M 1 mL	√	皮肤/眼刺激		关注		√	√		1	1	1	低	大量冲洗	
	Na₂HPO₄	—	0.1 M 1 mL	√	—		关注		√	√		1	1	1	低	大量冲洗	
	NaH₂PO₄	—	0.1 M 1 mL	√	—		关注		√	√		1	1	1	低	大量冲洗	
Ⅱ）在3支试管中各加入几滴 0.1 mol·L⁻¹ CaCl₂ 溶液，然后分别滴加 0.1 mol·L⁻¹ Na₃PO₄、0.1 mol·L⁻¹ Na₂HPO₄、0.1 mol·L⁻¹ NaH₂PO₄ 溶液，观察现象。	CaCl₂	—	0.1 M 几滴	√	—		关注		√			1	1	1	低	大量冲洗	
Ⅲ）在试管中滴加几滴 0.1 mol·L⁻¹ CuSO₄ 溶液，再滴加 0.5 mol·L⁻¹ Na₄P₂O₇ 溶液至过量，观察现象。	CuSO₄	—	0.1 M 几滴	√	皮肤/眼刺激		关注		√	√		1	1	1	低	大量冲洗	
	Na₄P₂O₇	—	0.5 M 20 mL	√	眼损伤		关注		√	√		1	1	1	低	大量冲洗	
Ⅳ）取1滴 0.1 mol·L⁻¹ CaCl₂ 溶液，滴加 0.1 mol·L⁻¹ Na₂CO₃ 溶液，再滴加 0.1 mol·L⁻¹ Na₅P₃O₁₀ 溶液，观察现象。	Na₂CO₃	—	0.1 M 几滴	√	眼刺激		关注		√	√		1	1	1	低	大量冲洗	
	Na₅P₃O₁₀	—	0.1 M 几滴	√	—		关注		√	√		1	1	1	低	大量冲洗	

续表

风险评估及控制工作表

实验任务						评估人	吴泽颖,王红松,吴星星			评估日期	2021年7月1日		
过程步骤		危害鉴别				健康风险	安全措施			评估清单/风险评估	风险评估		紧急情况/泼洒措施
		硼碳硅氮磷			化学品安全技术说明书	可能的伤害种类	降低或消除风险	技术	防护眼镜 手套 通风橱		风险评估	危害 可能性	风险	危害级别 高中低	特殊备注
化合物名称	类别	液度	用量	或物理危害									
H:磷酸根的测定													
取几滴 0.1 mol·L^{-1} NaPO$_3$ 溶液,加 0.5 mol·L^{-1} 浓硝酸,再加 1 mL 钼酸铵试剂,在水浴上微热到 40~45℃,观察现象。	NaPO$_3$	—	0.1 M 几滴		√	—	关注				1~4	1 1	低	大量冲洗
	HNO$_3$	8+ 5.1	0.5 M 几滴		√	皮肤灼伤和眼损伤	关注		√ √ √			1 3	3	低	大量冲洗
	钼酸铵试剂	—	1 mL		√	—	关注		√ √			1 1	1	低	大量冲洗
实验产物													
实验废弃物													
整个过程的最终评估	低风险(按照正确流程操作)												
签名	实验老师:					实验室主任/职业安全健康(OHS)代表:					学生/操作人:		

Experiment 7 Boron, carbon, silicon, nitrogen and phosphorus

【Objectives】

1. To master the important properties of boric acid and borax, understand the hydrolysis of soluble silicate and the formation and color of insoluble silicate.
2. To master the important properties of nitric acid, nitrite and their salts.
3. To understand the main properties of phosphate.
4. To learn and master the identification methods of CO_3^{2-}, NH_4^+, NO_2^-, NO_3^- and PO_4^{3-}.

【Principles】

Boric acid is a monobasic weak acid, and its dissociation in aqueous solution is different from other general meta-basic acid. Boric acid is one of Lewis acid that can react with polyhydric alcohols to increase the acidity of the solution.

Boric acid can be precipitated by reaction with acid. Borax solution can react with acid to form boric acid.

The CO_2 produced by reacting the carbonate solution with hydrochloric acid is passed into the $Ba(OH)_2$ solution, which can make the $Ba(OH)_2$ solution turbid. This method is used to identify CO_3^{2-}.

Most silicates are difficult to dissolve in water. Sodium silicate can hydrolyze.

There are two common methods to identify NH_4^+. One is using the reaction between NH_4^+ and OH^-. The generated $NH_3(g)$ turns the red litmus test paper to blue. The second is the reaction of NH_4^+ with Nessler reagent ($K_2[HgI_4]$ alkaline solution) to form a red-brown precipitate.

Nitrite is extremely unstable. The nitrite produced by the reaction of the nitrite solution with strong acid is decomposed into N_2O_3 and H_2O, and N_2O_3 can be decomposed into NO and NO_2.

The oxidation value of nitrogen in nitrite is +3. It acts as an oxidant in acidic solutions and is generally reduced to NO; when it reacts with a strong oxidant, it generates nitrate.

Nitric acid has strong oxidizability. It reacts with many non-metals and the main reduction product is NO. The reaction of concentrated nitric acid with metals mainly produces NO_2. The reaction of dilute nitric acid with metals usually generates NO. Active metals can reduce dilute nitric acid to NH_4^+.

The reaction of NO_2^- with $FeSO_4$ solution in HAc medium will form brown $[Fe(NO)(H_2O)_5]^{2+}$ (abbreviated as $[Fe(NO)]^{2+}$)

$$Fe^{2+} + NO_2^- + 2HAc \longrightarrow Fe^{3+} + NO + H_2O + 2Ac^-$$

$$Fe^{2+} + NO \longrightarrow [Fe(NO)]^{2+}$$

The reaction of NO_2^- with $FeSO_4$ solution in H_2SO_4 medium will form brown $[Fe(NO)]^{2+}$:

$$3Fe^{2+} + NO_3^- + 4H^+ \longrightarrow 3Fe^{3+} + NO + 2H_2O$$

$$Fe^{2+} + NO \longrightarrow [Fe(NO)]^{2+}$$

$[Fe(NO)]^{2+}$ formed at the interface between the test solution and the concentrated H_2SO_4 layer was a brown ring. This method is used to identify NO_3^- and is called the "brown ring" method. The presence of NO_2^- interferes the identification of NO_3^-. Adding urea and slightly heating can remove NO_3^-.

$$NO_2^- + CO(NH_2)_2 + 2H^+ \longrightarrow 2N_2 + CO_2 + 3H_2O$$

Alkali metal (except lithium), ammonium phosphate and monohydrogen phosphate are easily soluble in water, but other phosphates are hardly soluble in water. Most dihydrogen phosphates are easily soluble in water. Both pyrophosphate and tripolyphosphate have coordination effects.

PO_4^{3-} reacts with $(NH_4)_2MoO_4$ solution in a nitric acid medium to form a yellow ammonium phosphomolybdate precipitate. This reaction can be used to identify PO_4^{3-}.

【Apparatus and Chemicals】

1. Apparatus

Drip plate; water bath; pH test paper; red litmus test paper; nickel-chromium wire (one end is made into a ring).

2. Chemicals

HCl solution (2 mol·L^{-1}, 6 mol·L^{-1}, concentrated); H_2SO_4 (1 mol·L^{-1}, 6 mol·L^{-1}, concentrated); HNO_3 (2 mol·L^{-1}, concentrated); HAc (2 mol·L^{-1}); NaOH (2 mol·L^{-1}, 6 mol·L^{-1}); $Ba(OH)_2$ (saturated); Na_2CO_3 (0.1 mol·L^{-1}); $NaHCO_3$ (0.1 mol·L^{-1}); Na_2SiO_3 (0.5 mol·L^{-1}, 20%); NH_4Cl (0.1 mol·L^{-1}); $BaCl_2$ (0.5 mol·L^{-1}); $NaNO_2$ (0.1 mol·L^{-1}, 1 mol·L^{-1}); KI (0.02 mol·L^{-1}); $KMnO_4$ (0.01 mol·L^{-1}); KNO_3 (0.1 mol·L^{-1}); Na_3PO_4 (0.1 mol·L^{-1}); Na_2HPO_4 (0.1 mol·L^{-1}); NaH_2PO_4 (0.1 mol·L^{-1}); $CaCl_2$ (0.1 mol·L^{-1}); $CuSO_4$ (0.1 mol·L^{-1}); $Na_4P_2O_7$ (0.5 mol·L^{-1}); $Na_5P_3O_{10}$ (0.1 mol·L^{-1}); $Na_2B_4O_7 \cdot 10H_2O$(s); H_3BO_3(s); $Co(NO_3)_2 \cdot 6H_2O$(s); $CaCl_2$(s); $CuSO_4 \cdot 5H_2O$(s); $ZnSO_4 \cdot 7H_2O$(s); $Fe_2(SO_4)_3$(s); $NiSO_4 \cdot 7H_2O$(s); zinc powder; copper shavings; $FeSO_4 \cdot 7H_2O$(s); $Co(NH_2)_2$(s); NH_4NO_3(s); $Na_3PO_4 \cdot 12H_2O$(s); $NaHCO_3$(s); Na_2CO_3(s); glycerin; methyl orange indicator; Nessler reagent; starch test solution; ammonium molybdate reagent.

【Experimental Procedures】

1. Properties of Borax and Boric Acid

(1) Add about 0.5 g of boric acid crystals and 3 mL of deionized water to the test tube and observe the dissolution. After slightly warming, dissolve all, cool down to room temperature, and measure the pH of the solution with a pH test paper. Then add 1 drop of methyl orange indicator to

the solution, and divide the solution into two parts. Add 10 drops of glycerin into one part, mix well and compare the color of the two parts.

(2) Add about 1 g of borax and 2 mL of deionized water to the test tube, dissolve it with slight heat, and measure the pH of the solution with a pH test paper. Then add 1 mL of 6 mol · L^{-1} H$_2$SO$_4$ solution, cool the test tube in cold water, and continuously stir with a glass rod. After a while, observe the precipitation of boric acid crystals.

2. Identification of carbonate

Add 1 mL of 0.1 mol · L^{-1} Na$_2$CO$_3$ solution to the test tube, and then add a half drop tube of 2 mol · L^{-1} HCl solution in it. Cover the mouth with a stopper connecting with a tube immediately, and pass the generated gas into saturate Ba(OH)$_2$ solution. Observe the phenomenon.

3. Properties of silicate

1 mL of 0.5 mol · L^{-1} Na$_2$SiO$_3$ solution was added to a test tube, and its pH was measured with a pH test paper. Then add 6 mol · L^{-1} HCl solution dropwise to make the pH be 6~9, and observe the formation of silicic acid gel (if there is no gel, it can be slightly heated).

4. Ammonia test

(1) Add a small amount of 0.1 mol · L^{-1} NH$_4$Cl solution and 2 mol · L^{-1} NaOH solution to the test tube, and use a moist red litmus test paper to test the escaping gas at the mouth of the test tube. Write down the reaction equation

(2) Add 1 drop of Nessler reagent to the filter paper strip, and repeat the experiment (1) instead of the red litmus paper to observe the phenomenon.

5. Oxidation of nitric acid

(1) Put a small piece of copper shavings into the test tube, add a few drops of concentrated HNO$_3$, observe the phenomenon and then quickly dilute with water, discard the solution and recover the copper shavings.

(2) Put a small amount of zinc powder into the test tube, and add 1 mL 2 mol · L^{-1} HNO$_3$ to observe the phenomenon (if not reacting, it can be slightly warm).

6. Identification of NO$_3^-$ and NO$_2^-$

(1) Take 1 mL 0.1 mol · L^{-1} KNO$_3$ solution in a test tube, add a small amount of FeSO$_4$ · 7H$_2$O crystals and shake the test tube to dissolve it. Then hold the test tube obliquely, carefully add 1 mL of concentrated H$_2$SO$_4$ dropwise along the wall of the test tube, let it stand for a while, and observe the brown ring at the interface between the two liquids.

(2) Dilute 1 drop of 0.1 mol · L^{-1} NaNO$_2$ solution to 1 mL, add a small amount of FeSO$_4$ · 7H$_2$O crystals, shake the test tube to dissolve it, add 2 mol · L^{-1} HAc solution, and observe the phenomenon.

(3) Dilute 2 drops of 0.1 mol · L^{-1} KNO$_3$ solution and 2 drops of 0.1 mol · L^{-1} NaNO$_2$ solution to 1 mL, and add a small amount of urea and 2 drops of 1 mol · L^{-1} H$_2$SO$_4$ to eliminate the identification interference from NO$_2^-$, then perform a brown ring test.

7. Properties of phosphate

(1) The pH values of 0.1 mol · L^{-1} Na$_3$PO$_4$, 0.1 mol · L^{-1} Na$_2$HPO$_4$ and 0.1 mol · L^{-1} NaH$_2$PO$_4$ solutions were measured by pH test.

(2) Add several drops of 0.1 mol·L^{-1} CaCl$_2$ solution to each of the three test tubes, and then add 0.1 mol·L^{-1} Na$_3$PO$_4$, 0.1 mol·L^{-1} Na$_2$HPO$_4$ and 0.1 mol·L^{-1} NaH$_2$PO$_4$ solution respectively. Observe the phenomenon.

(3) Add a few drops of 0.1 mol·L^{-1} CuSO$_4$ solution to the test tube, and then add excess 0.5 mol·L^{-1} Na$_4$PO$_7$ solution dropwise, observe the phenomenon.

(4) Take 1 drop of 0.1 mol·L^{-1} CaCl$_2$ solution in test tube, add 0.1 mol·L^{-1} Na$_2$CO$_3$ solution dropwise, and then add 0.1 mol·L^{-1} Na$_5$P$_3$O$_{10}$ solution dropwise, observe the phenomenon.

8. Determination of phosphate

Take a few drops of 0.1 mol·L^{-1} NaPO$_3$ solution, add 0.5 mol·L^{-1} concentrated HNO$_3$, and then add 1 mL of ammonium molybdate reagent, heat slightly to 40~45 ℃ on the water bath, and then observe the phenomenon.

【Questions】

1. Why is boric acid different from common monobasic weak acid? Where does its acidity come from?

2. Why can addition of HAc solution, NH$_4$Cl solution or CO$_2$ to Na$_2$SiO$_3$ solution produce silicic acid gel?

3. When identifying NH$_4^+$, why drop the Nesler reagent on the filter paper to test the escaped NH$_3$ instead of adding the Nesler reagent directly to the solution containing NH$_4^+$?

4. What factors are involved in the main reduction products of the reaction of nitric acid and metals?

5. Does the presence of NO$_3^-$ interfere with the identification of NO$_2^-$?

6. Why use ammonium aluminate reagent to identify PO$_4^{3-}$ in nitric acid medium?

Risk Assessment and Control Worksheet

TASK	Boron, Carbon, Silicon, Nitrogen and Phosphorus				Assessor	Zeying WU, Hongsong WANG, Xingxing WU					Date	1 JUL. 2021	
Process Steps	Hazard Identification				Health Risk	Safety Measures					Risk Assessment		Action
	Chemical Name	Class	Concn.	Amount	MSDS	Type of Injury Possible	To reduce or eliminate the risk	Technique	Glasses	Gloves	Fume Cbd.	Checklist＼Risk Assessment	Risk Score	Risk Level	In case of Emergency/Spill
	or Physical Hazard											Likely 1~4	Hazard 1~4	H M L	Special Comments
A: Properties of Borax and Boric Acid															
Ⅰ) Add about 0.5 g of boric acid crystals and 3 mL of deionized water to the test tube and observe the dissolution. After slightly warming, dissolve all, cool down to room temperature, and measure the pH of the solution with a pH test paper. Then add 1 drop of methyl orange indicator to the solution, and divide the solution into two parts. Add 10 drops of glycerin into one part, mix well and compare the color of the two parts.	Boric acid crystals	—	—	0.5 g	√	—	Handle with care			√		1	1	1 L	Sponge & water
	Heating device					Electric shock & scald	Ensure cord is tagged	√	√			1	3	3 L	Report immediately to Demonstrator or Prep Room
Ⅱ) Add about 1 g of borax and 2 mL of deionized water to the test tube, dissolve it with slight heat, and measure the pH of the solution with a pH test paper. Then add 1 mL of 6 M H_2SO_4 solution, cool the test tube in cold water, and continuously stir with a glass rod. Add 10 drops of 1 M $NaNO_2$ solution to the test tube and then add 6 M H_2SO_4 solution dropwise.	Sulfuric acid	8	6 M	1 mL	√	Skin burns and eye injuries	Handle with care		√	√	√	1	1	1 L	Sponge & water

continued

Risk Assessment and Control Worksheet

TASK	Boron, Carbon, Silicon, Nitrogen and Phosphorus					Assessor	Zeying WU, Hongsong WANG, Xingxing WU				Date	1 JUL. 2021			
Process Steps	Hazard Identification					Health Risk	Safety Measures				Checklist～Risk Assessment			Action	
	Chemical Name	Class	Concn.	Amount	MSDS	Type of Injury Possible	To reduce or eliminate the risk	Technique	Glasses	Gloves	Fume Cbd.	Likely 1~4	Hazard 1~4	Risk Score	Risk Level H M L	In case of Emergency/Spill	Special Comments

B: Identification of carbonate

Process Steps	Chemical Name	Class	Concn.	Amount	MSDS	Type of Injury Possible	To reduce or eliminate the risk	Technique	Glasses	Gloves	Fume Cbd.	Likely	Hazard	Risk Score	Risk Level	In case of Emergency/Spill
Add 1 mL of 0.1 mol·L^{-1} Na$_2$CO$_3$ solution to the test tube, and then add a half drop tube of 2 mol·L^{-1} HCl solution in it. Cover the mouth with a stopper connecting with a tube immediately, and pass the generated gas into saturate Ba(OH)$_2$ solution. Observe the phenomenon.	Na$_2$CO$_3$	—	0.1 M	1 mL	√	eye irritant	Handle with care		√	√		1	1	1	L	Sponge & water
	HCl	2.3 +8	2 M	3 mL	√	Skin burns and eye injuries	care		√	√		1	3	3	L	Sponge & water
	Ba(OH)$_2$	8		20 mL	√	Skin burns and eye injuries	care		√	√		1	3	3	L	Sponge & water

C: Properties of silicate

Process Steps	Chemical Name	Class	Concn.	Amount	MSDS	Type of Injury Possible	To reduce or eliminate the risk	Technique	Glasses	Gloves	Fume Cbd.	Likely	Hazard	Risk Score	Risk Level	In case of Emergency/Spill
1 mL of 0.5 M Na$_2$SiO$_3$ solution was added to a test tube, and its pH was measured with a pH test paper. Then add 6 M HCl solution dropwise to make the pH be 6~9, and observe the formation of silicic acid gel.	Na$_2$SiO$_3$	—	0.5 M	1 mL	√	—	care		√	√	√	1	1	1	L	Sponge & water
	HCl	2.3 +8	6 M	1 mL	√	Skin burns and eye injuries	care		√	√	√	1	3	3	L	Sponge & water

D: Ammonia test

Process Steps	Chemical Name	Class	Concn.	Amount	MSDS	Type of Injury Possible	To reduce or eliminate the risk	Technique	Glasses	Gloves	Fume Cbd.	Likely	Hazard	Risk Score	Risk Level	In case of Emergency/Spill
I) Add a small amount of 0.1 mol·L^{-1} NH$_4$Cl solution and 2 mol·L^{-1} NaOH solution to the test tube, and use a moist red litmus test paper to test the escaping gas at the mouth of the test tube. Write down the reaction equation.	NH$_4$Cl	—	0.1 M	a little	√	eye irritant	care		√	√		1	2	2	L	Sponge & water
	NaOH	8	2 M	a little	√	Skin burns and eye injuries	care		√	√		1	3	3	L	Sponge & water

continued

Risk Assessment and Control Worksheet

TASK	Boron, Carbon, Silicon, Nitrogen and Phosphorus				Assessor	Zeying WU, Hongsong WANG, Xingxing WU					Date	1 JUL. 2021			
Process Steps	Hazard Identification				Health Risk		Safety Measures				Checklist～Risk Assessment	Risk Assessment			Action
	Chemical Name	Class	Concn.	Amount	MSDS	Type of Injury Possible	To reduce or eliminate the risk	Technique	Glasses	Gloves	Fume Cbd.		Likely	Hazard	Risk Score	Risk Level	In case of Emergency/Spill
		or Physical Hazard											1～4	1～4		H M L	Special Comments
II) Add 1 drop of Nessler reagent to the filter paper strip, and repeat the experiment (1) instead of the red litmus paper to observe the phenomenon.																	
E: Oxidation of nitric acid																	
I) Put a small piece of copper shavings into the test tube, add a few drops of concentrated HNO₃, observe the phenomenon and then quickly dilute with water, discard the solution and recover the copper shavings.	Copper shavings	—		a little	√	—				√			1	1	1	L	Sponge & water
	HNO₃	8+ 5.1	8 M	1 mL	√	Skin burns and eye injuries	care		√	√	√		1	3	3	L	Sponge & water
II) Put a small amount of zinc powder into the test tube, and add 1 mL 2 mol·L⁻¹ HNO₃ to observe the phenomenon (if not reacting, it can be slightly warm). Take the supernatant and check whether NH₄⁺ is generated.	Zinc powder	4.2+ 4.3		a little	√	—				√			1	1	1	L	Sponge & water
	HNO₃	8+ 5.1	2 M	1 mL	√	Skin burns and eye injuries	care		√	√	√		1	3	3	L	Sponge & water

continued

Risk Assessment and Control Worksheet

TASK	Boron, Carbon, Silicon, Nitrogen and Phosphorus				Assessor	Zeying WU, Hongsong WANG, Xingxing WU				Date	1 JUL. 2021	

Process Steps	Hazard Identification				Health Risk	Safety Measures				Checklist～Risk Assessment			Risk Assessment			Action
	Chemical Name	Class	Concn.	Amount	MSDS	Type of Injury Possible	To reduce or eliminate the risk	Technique	Glasses	Gloves	Fume Cbd.	Likely	Hazard	Risk Score	Risk Level	In case of Emergency/Spill Special Comments
		or Physical Hazard										$1\sim4$	$1\sim4$		H M L	

F: Identification of NO_3^- and NO_2^-

I) Take 1 mL 0.1 mol·L^{-1} KNO_3 solution in a test tube, add a small amount of $FeSO_4·7H_2O$ crystals and shake the test tube to dissolve it. Then hold the test tube obliquely, carefully add 1 mL of concentrated H_2SO_4 dropwise along the wall of the test tube, let it stand for a while, and observe the brown ring at the interface between the two liquids.

Chemical	Class	Concn.	Amount	MSDS	Injury	Reduce	Glasses	Gloves	Likely	Hazard	Score	Level	Action
KNO_3	5.1	0.1 M	1 mL	√	Skin & eye irritant	Handle with care				1	1	1	L	Sponge & water
$FeSO_4·7H_2O$ crystals	—	—	a little	√	—	Handle with care		√		1	1	1	L	Sponge & water
H_2SO_4	8	12 M	1 mL	√	Skin burns and eye injuries	Handle with care		√	√	√	1	4	4	L	Sponge & water

II) Dilute 1 drop of 0.1 mol·L^{-1} $NaNO_2$ solution to 1 mL, add a small amount of $FeSO_4·7H_2O$ crystals, shake the test tube to dissolve it, add 2 mol·L^{-1} HAc solution, and observe the phenomenon.

| HAc | 8+3 | 2 M | a little | √ | Skin burns and eye injuries | Handle with care | | √ | √ | √ | 1 | 1 | 1 | L | Sponge & water |

III) Dilute 2 drops of 0.1 mol·L^{-1} KNO_3 solution and 2 drops of 0.1 mol·L^{-1} $NaNO_2$ solution to 1 mL, and add a small amount of urea and 2 drops of 1 mol·L^{-1} H_2SO_4 to eliminate the identification interference from NO_2^-, then perform a brown ring test.

| Urea | — | — | a little | √ | — | Handle with care | | √ | √ | √ | 1 | 1 | 1 | L | Sponge & water |

Risk Assessment and Control Worksheet — continued

| TASK | Boron, Carbon, Silicon, Nitrogen and Phosphorus | Assessor | Zeying WU, Hongsong WANG, Xingxing WU | Date | 1 JUL. 2021 |

Process Steps	Hazard Identification				Health Risk		Safety Measures				Checklist〜Risk Assessment			Action	
	Chemical Name	Class	Concn.	Amount	MSDS	Type of Injury Possible	To reduce or eliminate the risk	Technique	Glasses	Gloves	Fume Cbd.	Likely $1\sim4$	Hazard $1\sim4$	Risk Score	Risk Level H M L	In case of Emergency/Spill / Special Comments
		or Physical Hazard														

G: Properties of phosphate

Process Steps	Chemical Name	Class	Concn.	Amount	MSDS	Type of Injury Possible	To reduce or eliminate the risk	Technique	Glasses	Gloves	Fume Cbd.	Likely	Hazard	Risk Score	Risk Level	Action
Ⅰ) The pH values of $0.1\ mol\cdot L^{-1}$ Na_3PO_4, $0.1\ mol\cdot L^{-1}$ Na_2HPO_4 and $0.1\ mol\cdot L^{-1}$ NaH_2PO_4 solutions were measured by pH test.	Na_3PO_4	—	0.1 M	1 mL	√	Skin&eye irritant	Handle with care		√	√		1	1	1	L	Sponge & water
	Na_2HPO_4	—	0.1 M	1 mL	√	—	Handle with care		√	√		1	1	1	L	Sponge & water
	NaH_2PO_4	—	0.1 M	1 mL	√	—	Handle with care		√	√		1	1	1	L	Sponge & water
Ⅱ) Add several drops of $0.1\ mol\cdot L^{-1}$ $CaCl_2$ solution to each of the three test tubes, and then add $0.1\ mol\cdot L^{-1}$ Na_3PO_4, $0.1\ mol\cdot L^{-1}$ Na_2HPO_4 and $0.1\ mol\cdot L^{-1}$ NaH_2PO_4 solution respectively. Observe the phenomenon.	$CaCl_2$	—	0.1 M	several drops	√	—	Handle with care		√	√		1	1	1	L	Sponge & water
Ⅲ) Add a few drops of $0.1\ mol\cdot L^{-1}$ $CuSO_4$ solution to the test tube, and then add excess $0.5\ mol\cdot L^{-1}$ Na_4PO_7 solution dropwise, observe the phenomenon.	$CuSO_4$	—	0.1 M	several drops	√	Skin&eye irritant	Handle with care		√	√		1	1	1	L	Sponge & water
	Na_4PO_7	—	0.5 M	20 mL	√	eye damage	Handle with care		√	√		1	1	1	L	Sponge & water

continued

Risk Assessment and Control Worksheet

TASK	Boron, Carbon, Silicon, Nitrogen and Phosphorus	Assessor	Zeying WU, Hongsong WANG, Xingxing WU	Date	1 JUL. 2021

Process Steps	Hazard Identification					Health Risk		Checklist — Risk Assessment						Risk Assessment			Action	
	Chemical Name	Class or Physical Hazard	Concn.	Amount	MSDS	Type of Injury Possible	To reduce or eliminate the risk	Technique	Glasses	Gloves	Fume Cbd.		Likely	Hazard	Risk Score	Risk Level H M L	In case of Emergency/Spill	Special Comments
IV) Take 1 drop of 0.1 mol·L⁻¹ CaCl₂ solution in test tube, add 0.1 mol·L⁻¹ Na₂CO₃ solution dropwise, and then add 0.1 mol·L⁻¹ Na₅P₃O₁₀ solution dropwise, observe the phenomenon.	Na₂CO₃	—	0.1 M	several drops	✓	Eye irritant	Handle with care		✓	✓		1~4	1~4	1	L	Sponge & water	
	Na₅P₃O₁₀	—	0.1 M	several drops	✓	—	Handle with care		✓	✓		1	1	1	L	Sponge & water	
H: Determination of phosphate																	
Take a few drops of 0.1 mol·L⁻¹ NaPO₃ solution, add 0.5 mol·L⁻¹ concentrated HNO₃, and then add 1 mL of ammonium molybdate reagent, heat slightly to 40~45 °C on the water bath, and then observe the phenomenon.	NaPO₃	—	0.1 M	several drops	✓	—	Handle with care		✓	✓		1	1	1	L	Sponge & water	
	HNO₃	8+ 5.1	0.5 M	several drops	✓	Skin burns and eye injuries	Handle with care		✓	✓		1	3	3	L	Sponge & water	
	Ammonium molybdate reagent	—		1 mL	✓	—	Handle with care		✓	✓		1	1	1	L	Sponge & water	
Experimental Product(s)																	
Experimental Waste(s)																	
Final assessment of overall process	Low risk when correct procedures followed																
Signatures	Supervisor:							Lab Manager/OHS Representative:						Student/Operator:				

实验 8　氧硫氯溴碘

【实验目的】

1. 掌握过氧化氢的还原性、亚硫酸及其盐的性质、硫代硫酸及其盐的性质和过二硫酸盐的氧化性。
2. 掌握卤素单质氧化性和卤化氢还原性的递变规律。
3. 掌握卤素含氧酸盐的氧化性。
4. 学会 H_2O_2、S^{2-}、SO_3^{2-}、$S_2O_3^{2-}$、Cl^-、Br^-、I^- 的鉴定方法。

【实验原理】

过氧化氢具有强氧化性。它也能被更强的氧化剂氧化为氧气。

H_2O_2 的鉴定：酸性溶液中，H_2O_2 与 $Cr_2O_7^{2-}$ 反应生成蓝色的 CrO_5。

S^{2-} 的鉴定：一般有两种方法，一是在含有 S^{2-} 的溶液中加入稀盐酸，由于生成的 H_2S 具有强还原性，能使湿润的 $Pb(Ac)_2$ 试纸变黑。二是在碱性溶液中，S^{2-} 与 $[Fe(CN)_5NO]^{2-}$ 反应生成紫色配合物。

$$S^{2-} + [Fe(CN)_5NO]^{2-} \longrightarrow [Fe(CN)_5NOS]^{4-}$$

SO_2 溶于水生成不稳定的亚硫酸。亚硫酸及其盐常用作还原剂，但遇到强还原剂时也起氧化作用。H_2SO_3 可与某些有机物发生加成反应生成无色加成物，所以具有漂白性。

SO_3^{2-} 的鉴定：SO_3^{2-} 与 $[Fe(CN)_5NO]^{2-}$ 反应生成红色配合物，加入饱和 $ZnSO_4$ 溶液和 $K_4[Fe(CN)_6]$ 溶液，会使红色明显加深。

硫代硫酸不稳定，其盐遇酸容易分解。$Na_2S_2O_3$ 常用作还原剂，还能与某些金属离子形成配合物。

$S_2O_3^{2-}$ 的鉴定：$S_2O_3^{2-}$ 与 Ag^+ 反应能生成白色的 $Ag_2S_2O_3$ 沉淀，而 $Ag_2S_2O_3(s)$ 能迅速分解为 Ag_2S 和 H_2SO_4，悬浊液颜色由白色变为黄色、棕色，最后变为黑色。涉及的反应如下：

$$2Ag^+ + S_2O_3^{2-} \longrightarrow Ag_2S_2O_3(s)$$

$$Ag_2S_2O_3(s) + H_2O \longrightarrow Ag_2S(s) + H_2SO_4$$

过二硫酸盐是强氧化剂，在酸性条件下能将 Mn^{2+} 氧化为 MnO_4^-，Ag^+ 可催化此反应。

氯、溴、碘氧化性的强弱次序为 $Cl_2 > Br_2 > I_2$。卤化氢还原性强弱的次序为 $HI > HBr > HCl$。HBr 和 HI 能分别将浓硫酸还原为 SO_2 和 H_2S。Br^- 能被 Cl_2 氧化为 Br_2，在 CCl_4 中呈棕黄色。I^- 能被 Cl_2 氧化为 I_2，在 CCl_4 中呈紫色；当 Cl_2 过量时，I_2 被氧化为无色的 IO_3^-。

次氯酸及其盐具有强氧化性。卤酸盐在酸性条件下都具有强氧化性，其强弱次序为 $BrO_3^- > ClO_3^- > IO_3^-$。

Cl^-、Br^-、I^- 与 Ag^+ 反应分别生成 $AgCl$、$AgBr$、AgI 沉淀，它们的溶度积依次减小，都不溶于稀硝酸。

Cl^- 的鉴定：$AgCl$ 能溶于稀氨水或 $(NH_4)_2CO_3$ 溶液，生成 $[Ag(NH_3)_2]^+$，再加入稀硝酸时，$AgCl$ 会重新沉淀出来。

Br^- 和 I^- 的鉴定：AgBr 和 AgI 不溶于稀氨水或 $(NH_4)_2CO_3$ 溶液，在 HAc 介质中能被锌还原为 Ag 单质，而 Br^- 和 I^- 可用氯水将其氧化。

【器材与试剂】

1. 器材：点滴板，pH 试纸，淀粉-KI 试纸，$Pb(Ac)_2$ 试纸，蓝色石蕊试纸。

2. 试剂：H_2SO_4（1 mol·L^{-1}，2 mol·L^{-1}，1+1，浓），HCl 溶液（2 mol·L^{-1}，浓），HNO_3（2 mol·L^{-1}，浓），HAc（6 mol·L^{-1}），NaOH（2 mol·L^{-1}），$NH_3 \cdot H_2O$（2 mol·L^{-1}），KI（0.1 mol·L^{-1}），KBr（0.1 mol·L^{-1}，0.5 mol·L^{-1}），$K_2Cr_2O_7$（0.1 mol·L^{-1}），NaCl（0.1 mol·L^{-1}），$KMnO_4$（0.01 mol·L^{-1}），$KClO_3$（饱和），$KBrO_3$（饱和），KIO_3（0.1 mol·L^{-1}），$FeCl_3$（0.1 mol·L^{-1}），$ZnSO_4$（饱和），$Na_2[Fe(CN)_5NO]$（1%），$K_4[Fe(CN)_6]$（0.1 mol·L^{-1}），$Na_2S_2O_3$（0.1 mol·L^{-1}），Na_2SO_3（0.1 mol·L^{-1}），Na_2S（0.1 mol·L^{-1}），$(NH_4)_2CO_3$（12%），$AgNO_3$（0.1 mol·L^{-1}），$(NH_4)_2S_2O_8$（0.2 mol·L^{-1}），$BaCl_2$（1 mol·L^{-1}），$MnSO_4$（0.1 mol·L^{-1}），$NaHSO_3$（0.1 mol·L^{-1}），MnO_2(s)，$(NH_4)_2S_2O_8$(s)，硫粉，NaCl(s)，KBr(s)，KI(s)，锌粒，CCl_4，戊醇，H_2S 溶液（饱和），SO_2 溶液（饱和），H_2O_2（3%），碘水（0.01 mol·L^{-1}，饱和），淀粉试液，品红溶液，氯水（饱和）。

【实验内容】

1. 过氧化氢的性质

在试管中加入 3% 的 H_2O_2 溶液和戊醇各 0.5 mL，加几滴 1 mol·L^{-1} H_2SO_4 溶液和 1 滴 0.1 mol·L^{-1} $K_2Cr_2O_7$ 摇荡试管，观察并解释现象。

2. 硫代硫酸及其盐的性质

（1）加入几滴 0.1 mol·L^{-1} $Na_2S_2O_3$ 溶液和 2 mol·L^{-1} HCl 溶液于试管中，摇荡片刻，观察现象，用湿润的蓝色石蕊试纸检验逸出的气体。写出反应的化学方程式。

（2）试管中加几滴 0.01 mol·L^{-1} 碘水，加 1 滴淀粉试液，再逐滴加入 0.1 mol·L^{-1} $Na_2S_2O_3$ 溶液，观察并用化学方程式解释观察到的现象。

（3）在点滴板上滴 1 滴 0.1 mol·L^{-1} $Na_2S_2O_3$ 溶液，再滴加 0.1 mol·L^{-1} $AgNO_3$ 溶液至生成白色沉淀，观察颜色的变化。解释现象。

3. 氯、溴、碘含氧酸盐的氧化性

（1）试管中取几滴饱和 $KClO_3$ 溶液，加入几滴浓盐酸，检验逸出的气体。写出反应的化学方程式。

（2）取 2~3 滴 0.1 mol·L^{-1} KI 溶液，加入 4 滴饱和 $KClO_3$ 溶液，再逐滴加入 1∶1 的 H_2SO_4 溶液，不断摇荡，观察溶液颜色的变化并用化学方程式解释现象。

（3）取几滴 0.1 mol·L^{-1} $KClO_3$ 溶液，加入几滴稀硫酸酸化后加数滴 CCl_4，再滴加 0.1 mol·L^{-1} Na_2HSO_3 溶液，摇荡并观察现象。写出反应的离子方程式。

4. Cl^-、Br^- 和 I^- 的鉴定

在试管中加 2 滴 0.1 mol·L^{-1} NaCl 溶液、1 滴 2 mol·L^{-1} HNO_3 溶液和 2 滴 0.1 mol·L^{-1} $AgNO_3$ 溶液，振荡并观察现象。加入数滴 2 mol·L^{-1} 氨水，摇荡使沉淀溶解，再加入 2 mol·L^{-1} HNO_3 溶液，观察有何变化。写出有关反应的离子方程式。

【思考题】

1. 实验室长期放置的 H_2S 溶液、Na_2S 溶液和 Na_2SO_3 溶液会发生什么变化?
2. 鉴定 $S_2O_3^{2-}$ 时,$AgNO_3$ 溶液应过量,否则会出现什么现象? 为什么?
3. 鉴定 Cl^- 时,为什么要先加稀硝酸? 而鉴定 Br^- 和 I^- 时为什么先加稀硫酸而不加稀硝酸?

● 视频演示

风险评估及控制工作表

评估日期：2021年7月1日
评估人：吴泽颖，王红松，吴星星
实验任务：氧硫氯溴碘

实验任务 过程步骤	化合物名称	危害鉴别 类别	浓度	用量	化学品安全技术说明书	健康风险 可能的伤害种类	降低或消除风险	安全措施 技术	防护眼镜	手套	通风橱	风险评估 可能性 1~4	危害 1~4	风险	危害级别 高中低	紧急情况措施	特殊备注
A：过氧化氢的性质																	
取3%的H₂O₂的溶液和戊醇各0.5 mL，加入1 mol·L⁻¹ H₂SO₄溶液和1 mol·L⁻¹ K₂Cr₂O₇溶液至1滴，摇荡试管，观察现象。	H₂O₂	5.1+8	3%	0.5 mL	√	皮肤灼伤和眼损伤	关注		√	√		1	3	3	低	大量冲洗	
	戊醇	3	—	0.5 mL	√	皮肤/呼吸道刺激	关注		√	√		1	2	2	低	大量冲洗	
	硫酸	8	1 M	几滴	√	皮肤灼伤和眼损伤	关注		√	√		1	4	4	低	大量冲洗	
	K₂Cr₂O₇	5.1+6.1	0.1 M	1滴	√	皮肤灼伤和眼损伤	关注		√	√		1	2	2	低	大量冲洗	
B：硫代硫酸及其盐的性质																	
Ⅰ) 在试管中加入几滴0.1 mol·L⁻¹ Na₂S₂O₃溶液和2 mol·L⁻¹ HCl溶液，摇荡片刻，观察现象。	Na₂S₂O₃溶液	—	0.1 M	几滴	√	—	关注		√	√	√	1	1	1	低	大量冲洗	
	盐酸	2.3+8	2 M	1 mL	√	皮肤灼伤和眼损伤	关注		√	√	√	1	3	3	低	大量冲洗	
Ⅱ) 取几滴0.01 mol·L⁻¹碘水，加1滴淀粉试液，逐滴加入0.1 mol·L⁻¹ Na₂S₂O₃溶液，观察现象。	碘水	—	0.01 M	几滴	√	皮肤接触有害	关注		√	√	√	1	3	3	低	大量冲洗	
Ⅲ) 在点滴板上加1滴0.1 mol·L⁻¹ Na₂S₂O₃溶液，再滴加0.1 mol·L⁻¹ AgNO₃溶液至沉淀生成白色沉淀，观察颜色的变化。	AgNO₃	5.1	0.1 M	几滴	√	皮肤灼伤和眼损伤	关注		√	√	√	1	3	3	低	大量冲洗	
C：氯、溴、碘含氧酸盐的氧化性																	
Ⅰ) 取几滴饱和KClO₃溶液，加入几滴浓盐酸，并检验逸出的气体。	盐酸	2.3+8	12 M	1 mL	√	皮肤灼伤和眼损伤	关注		√	√		1	3	3	低	大量冲洗	
	饱和KClO₃溶液	5.1		几滴	√	皮肤灼伤和眼损伤	关注		√	√		1	4	4	低	大量冲洗	

续表

风险评估及控制工作表

实验任务	过程步骤	化合物名称	危害鉴别			化学品安全技术说明书	健康风险		安全措施				风险评估			紧急情况措施	特殊备注
			类别	浓度	用量		可能的伤害种类	降低或消除风险	技术	防护眼镜	手套	通风橱	可能性	危害	风险		
			或物理危害										1～4	1～4	1～4		
	Ⅱ）取 2~3 滴 0.1 mol·L⁻¹ KI 溶液，加入 4 滴饱和 KClO₃ 溶液，再逐滴加入 H₂SO₄(1+1)，不断摇荡，观察溶液颜色的变化。	KI	—	0.1 M	2～3滴	√	—	关注		√	√	√	1	1	1	大量冲洗	
		饱和 KClO₃ 溶液	5.1		4滴	√	—	关注		√	√	√	1	1	1	大量冲洗	
		H₂SO₄(1+1) 溶液	8		几滴	√	皮肤灼伤和眼睛损伤	关注		√	√	√	1	4	4	大量冲洗	
	Ⅲ）取几滴 0.1 mol·L⁻¹ KClO₃ 溶液，酸化后加数滴 CCl₄，再滴加 0.1 mol·L⁻¹ Na₂HSO₃ 溶液，摇荡，观察现象。	KClO₃	5.1	0.1 M	几滴	√	—	关注		√	√	√	1	1	1	大量冲洗	
		CCl₄	6.1		几滴	√	皮肤接触会中毒	关注		√	√	√	1	4	4	大量冲洗	
		Na₂HSO₃	—	0.1 M	几滴	√	皮肤刺激	关注		√	√	√	1	1	1	大量冲洗	
D: Cl⁻、Br⁻ 和 I⁻ 的鉴定	取 2 滴 0.1 mol·L⁻¹ NaCl 溶液，加入 2 mol·L⁻¹ HNO₃ 溶液和 2 滴 0.1 mol·L⁻¹ AgNO₃ 溶液，观察现象。在沉淀中加入数滴 2 mol·L⁻¹ 氨水溶液，再加数滴 2 mol·L⁻¹ HNO₃ 溶液，观察有何变化。	NaCl	—	0.1 M	2滴	√	—	关注		√	√	√	1	1	1	大量冲洗	
		硝酸	8+5.1	2 M	1滴	√	皮肤灼伤和眼睛损伤	关注		√	√	√	1	3	3	大量冲洗	
		AgNO₃	5.1	0.1 M	2滴	√	皮肤灼伤和眼睛损伤	关注		√	√	√	1	3	3	大量冲洗	
		氨水	8	2 M	几滴	√	皮肤灼伤和眼睛损伤	关注		√	√	√	1	3	3	大量冲洗	
实验产物																	
实验废弃物																	
整个过程的最终评估	低风险（按照正确流程操作）																

评估清单　评估人　吴泽颖，王红松，吴星星　评估日期　2021年7月1日

签名　　实验老师：　　　　实验室主任 职业安全健康(OHS)代表：　　　　学生/操作人：

Experiment 8　Oxygen, sulfur, chlorine, bromine and iodine

【Objectives】

1. To master the reducibility of hydrogen peroxide, the properties of sulfite and its salts, the properties of thiosulfate and its salts, and the oxidation of persulfate.
2. To master the gradual change rule of halogen element oxidizing property and hydrogen halide reducing property.
3. To master the oxidizing property of halogen oxyacid salt.
4. To Learn the identification methods of H_2O_2、S^{2-}、SO_3^{2-}、$S_2O_3^{2-}$、Cl^-、Br^- and I^-。

【Principles】

Hydrogen peroxide has strong oxidizability. It can also be oxidized into oxygen by a stronger oxidant.

In acidic solution, H_2O_2 can reacts with $Cr_2O_7^{2-}$ to produce blue CrO_5. This reaction is used to identify H_2O_2.

H_2S has strong reducibility. Add dilute hydrochloric acid to the solution containing S^{2-}, the generated H_2S gas can make the wet $Pb(Ac)_2$ test paper black. In alkaline solution, S^{2-} reacts with $[Fe(CN)_5NO]^{2-}$ to form a purple complex:

$$S^{2-} + [Fe(CN)_5NO]^{2-} \longrightarrow [Fe(CN)_5NOS]^{4-}$$

These two methods are used to identify S.

SO_2 dissolves in water to generate unstable sulfurous acid. Sulfurous acid and its salts are often used as reducing agents, but they also can be oxidized when encountering strong reducing agents. H_2SO_4 can react with some organic substances to form colorless adducts, which give it bleaching properties, and these adducts tend to decompose when heated. SO_3^{2-} reacts with $[Fe(CN)_5NO]^{2-}$ to form a red complex, Adding saturated $ZnSO_4$ solution and $K_4[Fe(CN)_6]$ solution will deepen the red color obviously. This method is used to identify SO_3^{2-}.

Thiosulfuric acid is unstable and its salt is easily decomposed when it encounters acid. $Na_2S_2O_3$ is often used as a reducing agent and can also form complexes with certain metal ions. $S_2O_3^{2-}$ reacts with Ag^+ to produce a white $Ag_2S_2O_3$ precipitate:

$$2Ag^+ + S_2O_3^{2-} \longrightarrow Ag_2S_2O_3(s)$$

$Ag_2S_2O_3(s)$ can be quickly decomposed into Ag_2S and H_2SO_4.

$$Ag_2S_2O_3(s) + H_2O \longrightarrow Ag_2S(s) + H_2SO_4$$

This process is accompanied by the color changing from white to yellow or brown, and finally to black. This method is used to identify $S_2O_3^{2-}$.

Peroxodisulfate is a strong oxidant. it can oxidize Mn^{2+} into MnO_4^- under acidic conditions, the reaction rate increases in the presence of Ag^+ (as a catalyst).

The order of oxidizing strength of chlorine, bromine and iodine is $Cl_2 > Br_2 > I_2$. The order of reducing strength of hydrogen halide is $HI > HBr > HCl$. HBr and HI can respectively reduce concentrated H_2SO_4 to SO_2 and H_2S. Br^- can be oxidized by Cl_2 to Br_2, which is brownish in CCl_4. I^- can be oxidized by Cl_2 to I_2, which is purple in CCl_4; when Cl_2 is excessive, I_2 is oxidized to colorless IO_3^-.

Hypochlorous acid and its salts have strong oxidizing properties. Under acidic conditions, halide salts have strong oxidizing properties, and the order of their strength is $BrO_3^- > ClO_3^- > IO_3^-$.

Cl^-, Br^-, I^- respectively react with Ag^+ to form AgCl, AgBr, and AgI precipitates, their solubility product constant decrease in sequence, and they are not soluble in dilute HNO_3. AgCl can dissolve in dilute ammonia water or $(NH_4)_2CO_3$ solution to generate $[Ag(NH_3)_2]^+$. When dilute HNO_3 is added, AgCl will re-precipitate, which can identify the presence of Cl^-. AgBr and AgI are insoluble in dilute ammonia water or $(NH_4)_2CO_3$ solution. Br^- and I^- can be transferred into the solution, and then oxidized with chlorine water to identify the presence of Br^- and I^-.

【Apparatus and Chemicals】

1. Apparatus

Drip plate; pH test paper; starch-KI test paper; $Pb(Ac)_2$ test paper; blue litmus test paper.

2. Chemicals

H_2SO_4 (1 mol·L^{-1}, 2 mol·L^{-1}, 1+1, concentrated); HCl 溶液 (2 mol·L^{-1}, concentrated); HNO_3 (2 mol·L^{-1}, concentrated); HAc (6 mol·L^{-1}); NaOH (2 mol·L^{-1}); $NH_3·H_2O$ (2 mol·L^{-1}); KI (0.1 mol·L^{-1}); KBr (0.1 mol·L^{-1}, 0.5 mol·L^{-1}); $K_2Cr_2O_7$ (0.1 mol·L^{-1}); NaCl (0.1 mol·L^{-1}); $KMnO_4$ (0.01 mol·L^{-1}); $KClO_3$ (saturated); $KBrO_3$ (saturated); KIO_3 (0.1 mol·L^{-1}); $FeCl_3$ (0.1 mol·L^{-1}); $ZnSO_4$ (saturated); $Na_2[Fe(CN)_5NO]$ (1%); $K_4[Fe(CN)_6]$ (0.1 mol·L^{-1}); $Na_2S_2O_3$ (0.1 mol·L^{-1}); Na_2SO_3 (0.1 mol·L^{-1}); Na_2S (0.1 mol·L^{-1}); $(NH_4)_2CO_3$ (12%); $AgNO_3$ (0.1 mol·L^{-1}); $(NH_4)_2S_2O_8$ (0.2 mol·L^{-1}); $BaCl_2$ (1 mol·L^{-1}); $MnSO_4$ (0.1 mol·L^{-1}); $NaHSO_3$ (0.1 mol·L^{-1}), MnO_2(s); $(NH_4)_2S_2O_8$(s); sulfur powder; NaCl(s), KBr(s); KI(s); zinc particles; CCl_4; pentanol; H_2S solution (saturated); SO_2 solution (saturated); H_2O_2 (3%); iodine water (0.01 mol·L^{-1}, saturated); starch test solution; magenta solution; chlorine water (saturated).

【Experimental Procedures】

1. The property of hydrogen peroxide

Take 0.5 mL of 3% H_2O_2 solution amyl alcohol, add a few drops of 1 mol·L^{-1} H_2SO_4 solution and 1 drop of 0.1 mol·L^{-1} $K_2Cr_2O_7$ and shake the test tube. Observe the phenomenon and write the reaction equation.

2. Properties of thiosulfuric acid and its salts

(1) Add a few drops of 0.1 mol·L^{-1} Na$_2$S$_2$O$_3$ solution and 2 mol·L^{-1} HCl solution to the test tube, shake it for a while. Observe the phenomenon and check the escaped gas with the wet blue litmus paper. Write the reaction equation.

(2) Take a few drops of 0.01 mol·L^{-1} iodine water, add 1 drop of starch test solution, and add 0.1 mol·L^{-1} Na$_2$S$_2$O$_3$ solution drop by drop. Observe the phenomenon and write down the reaction equation.

(3) Add 1 drop of 0.1 mol·L^{-1} Na$_2$S$_2$O$_3$ solution to the drop plate, then add 0.1 mol·L^{-1} AgNO$_3$ solution dropwise until a white precipitate is formed. Observe the color change and write down the relevant reaction equations.

3. Oxidizing properties of chlorine, bromine and iodine oxygenates

(1) Take a few drops of saturated KClO$_3$ solution, add a few drops of concentrated hydrochloric acid, and check the escaping gas. Write the reaction equation.

(2) Take 2~3 drops of 0.1 mol·L^{-1} KI solution, add 4 drops of saturated KClO$_3$ solution, then add H$_2$SO$_4$(1+1) solution drop by drop, shake it continuously and observe the color change of the solution. Write down the reaction equation for each step.

(3) Take a few drops of 0.1 mol·L^{-1} KClO$_3$ solution, add a few drops of CCl$_4$ after acidification, then add 0.1 mol·L^{-1} Na$_2$HSO$_3$ solution, shake it and observe the phenomenon. Write down the ion reaction equation.

4. Identification of Cl$^-$、Br$^-$ 和 I$^-$

Take 2 drops of 0.1 mol·L^{-1} NaCl solution, add 1 drop of 2 mol·L^{-1} HNO$_3$ solution and 0.1 mol·L^{-1} AgNO$_3$ solution. Observe the phenomenon. Add a few drops of 2 mol·L^{-1} ammonia solution into the precipitate, shake to dissolve the precipitate, and then add a few drops of 2 mol·L^{-1} HNO$_3$ solution to observe any changes. Write down the ion reaction equation.

【Questions】

1. What will happen to the H$_2$S solution, Na$_2$S solution and Na$_2$SO$_3$ solution which placed in the laboratory for a long time?

2. When identifying S$_2$O$_3^{2-}$, the AgNO$_3$ solution should be excessive, otherwise what will happen? why?

3. Why dilute HNO$_3$ should be added when identifying Cl$^-$? Why dilute H$_2$SO$_4$ instead of dilute HNO$_3$ should be added when identifying Br$^-$ and I$^-$?

Risk Assessment and Control Worksheet

| TASK | Oxygen, Sulfur, Chlorine, Bromine and Iodine | Assessor | Zeying WU, Hongsong WANG, Xingxing WU | Date | 1 JUL. 2021 |

Process Steps	Hazard Identification				Health Risk		Safety Measures				Checklist～Risk Assessment			Action	
	Chemical Name	Class	Concn.	Amount	MSDS	Type of Injury Possible	To reduce or eliminate the risk	Technique	Glasses	Gloves	Fume Cbd.	Likely 1~4	Hazard 1~4	Risk Score	Risk Level H M L	In case of Emergency/Spill Special Comments

A: The property of hydrogen peroxide

		or Physical Hazard														
Take 0.5 mL of 3% H_2O_2 solution and amyl alcohol, add a few drops of 1 M H_2SO_4 solution and 1 drop of 0.1 M $K_2Cr_2O_7$ and shake the test tube. Observe the phenomenon and write the reaction equation.	H_2O_2 solution	5.1+8	3%	0.5 mL	√	Skin burns and eye injuries	Handle with care		√	√		1	3	3	L	Sponge & water
	Amyl alcohol	3		0.5 mL	√	skin and respiratory irritant	Handle with care		√	√	√	1	2	2	L	Sponge & water
	H_2SO_4	8	1 M	several drops	√	Skin burns and eye injuries	Handle with care		√	√		1	4	4	L	Sponge & water
	$K_2Cr_2O_7$	5.1+6.1	0.1 M	1 drop	√	Skin burns and eye injuries	Handle with care		√	√		1	2	2	L	Sponge & water

B: Properties of thiosulfuric acid and its salts

I) Add a few drops of 0.1 mol·L⁻¹ $Na_2S_2O_3$ solution and 2 mol·L⁻¹ HCl solution to the test tube, shake it for a while.	$Na_2S_2O_3$	—	0.1 M	several drops	√	—	Handle with care		√	√	√	1	1	1	L	Sponge & water
	HCl	2,3+8	2 M	1 mL	√	Skin burns and eye injuries	Handle with care		√	√	√	1	3	3	L	Sponge & water
II) Take a few drops of 0.01 mol·L⁻¹ iodine solution, add 1 drop of starch test solution, and add 0.1 mol·L⁻¹ $Na_2S_2O_3$ solution drop by drop. Observe the phenomenon and write down the reaction equation.	Iodine solution	—	0.01 M	several drops	√	Harmful by skin contact	Handle with care		√	√		1	3	3	L	Sponge & water

Risk Assessment and Control Worksheet

TASK: Oxygen, Sulfur, Chlorine, Bromine and Iodine **Assessor:** Zeying WU, Hongsong WANG, Xingxing WU **Date:** 1 JUL. 2021

Process Steps	Hazard Identification				Health Risk	To reduce or eliminate the risk	Safety Measures				Checklist＼Risk Assessment			Action		
	Chemical Name	Class	Concn.	Amount	MSDS / or Physical Hazard	Type of Injury Possible		Technique	Glasses	Gloves	Fume Cbd.	Likely 1~4	Hazard 1~4	Risk Score	Risk Level H M L	In case of Emergency/Spill / Special Comments
Ⅲ) Add 1 drop of 0.1 mol·L⁻¹ Na₂S₂O₃ solution to the drop plate, then add 0.1 mol·L⁻¹ AgNO₃ solution dropwise until a white precipitate is formed. Observe the color change and write down the relevant reaction equations.	AgNO₃	5.1	0.1 M	several drops	✓	Skin burns and eye injuries	Handle with care		✓	✓		1	3	3	L	Sponge & water
C: Oxidizing properties of chlorine, bromine and iodine oxygenates															
Ⅰ) Take a few drops of saturated KClO₃ solution, add a few drops of concentrated hydrochloric acid, and check the escaping gas. Write the reaction equation.	HCl	2.3 +8	12 M	1 mL	✓	Skin burns and eye injuries	Handle with care		✓	✓	✓	1	3	3	L	Sponge & water
	Saturated KClO₃ solution	5.1		several drops	✓	—	Handle with care		✓	✓	✓	1	4	4	L	Sponge & water
Ⅱ) Take 2~3 drops of 0.1 mol·L⁻¹ KI solution, add 4 drops of saturated KClO₃ solution, then add H₂SO₄ (1+1) solution drop by drop, shake it continuously and observe the color change of the solution. Write down the reaction equation for each step.	KI		0.1 M	2~3 drops	✓	—	Handle with care		✓	✓	✓	1	1	1	L	Sponge & water
	Saturated KClO₃ solution	5.1		4 drops	✓	—	Handle with care		✓	✓	✓	1	1	1	L	Sponge & water
	H₂SO₄	8		several drops	✓	Skin burns and eye injuries	Handle with care		✓	✓	✓	1	4	4	L	Sponge & water

Risk Assessment and Control Worksheet

TASK	Oxygen, Sulfur, Chlorine, Bromine and Iodine			Assessor	Zeying WU, Hongsong WANG, Xingxing WU					Date	1 JUL. 2021			Action	
Process Steps	Hazard Identification			Health Risk	Safety Measures				Checklist～Risk Assessment				In case of Emergency/Spill	
	Chemical Name	Class	Concn.	Amount	MSDS	Type of Injury Possible	To reduce or eliminate the risk	Technique	Glasses	Gloves	Fume Cbd.	Likely	Hazard	Risk Score	Risk Level	
		or Physical Hazard										1~4	1~4		H M L	Special Comments
Ⅲ) Take a few drops of 0.1 mol·L^{-1} KClO$_3$ solution, add a few drops of CCl$_4$ after acidification, then add 0.1 mol·L^{-1} Na$_2$HSO$_3$ solution, shake it and observe the phenomenon. Write down the ion reaction equation.	KClO$_3$	5.1	0.1 M	several drops	√	—	Handle with care					1	1	1	L	Sponge & water
	Na$_2$HSO$_3$	—	0.1 M	several drops	√	Skin irritant	Handle with care		√	√		1	1	1	L	Sponge & water
	CCl$_4$	6.1		several drops	√	Skin contact can be toxic	Handle with care					1	4	4	L	Sponge & water
D: Identification of Cl$^-$, Br$^-$ 和 I$^-$																
Take 2 drops of 0.1 mol·L^{-1} NaCl solution, add 1 drop of 2 mol·L^{-1} HNO$_3$ solution and 0.1 mol·L^{-1} AgNO$_3$ solution. Observe the phenomenon. Add a few drops of 2 mol·L^{-1} ammonia solution into the precipitate, shake to dissolve the precipitate, and then add a few drops of 2 mol·L^{-1} HNO$_3$ solution to observe any changes. Write down the ion reaction equation.	NaCl	—	0.1 M	2 drops	√	—	Handle with care		√	√	√	1	1	1	L	Sponge & water
	Nitric acid	8+ 5.1	2 M	1 drop	√	Skin burns and eye injuries	Handle with care		√	√	√	1	3	3	L	Sponge & water
	AgNO$_3$	5.1	0.1 M	2 drops	√	Skin burns and eye injuries	Handle with care		√	√	√	1	3	3	L	Sponge & water
	Ammonia	8	2 M	several drops	√	Skin burns and eye injuries	Handle with care		√	√	√	1	3	3	L	Sponge & water

Experimental Product (s)

Experimental Waste (s)

Final assessment of overall process   Low risk when correct procedures followed

Signatures   Supervisor: Lab Manager/OHS Representative: Student/Operator:

实验 9　钛、钒、铬和锰的化合物

【实验目的】

1. 掌握钛、钒、铬和锰的某些重要化合物（氢氧化物、盐类、配位化合物等）的性质。
2. 掌握钛、钒、铬和锰的氧化还原性。

【实验原理】

1. 钛、钒、铬、锰在元素周期表中的位置

钛、钒、铬、锰在元素周期表中的位置如表 9-1 所示。

表 9-1　钛、钒、铬、锰在元素周期表中的位置

周期 \ 族	ⅣB	ⅤB	ⅥB	ⅦB
	钛分族	钒分族	铬分族	锰分族
4	22 Ti [Ar]$3d^2 4s^2$	23 V [Ar]$3d^3 4s^2$	24 Cr [Ar]$3d^5 4s^1$	25 Mn [Ar]$3d^5 4s^2$

2. 氢氧化物和盐类

这些元素的氢氧化物的酸碱性符合一般规律。M(OH)$_2$ 都是难溶的中强碱；M(OH)$_3$ 是弱碱且酸性依次增强；Cr(OH)$_3$ 是典型的两性氢氧化物。Cr(Ⅲ)既能形成阳离子盐也能形成阴离子盐，高氧化值时，亲氧性很强，因此即使在强酸性溶液中也不存在简单阳离子（水合）Ti^{4+}、V^{4+}、V^{5+}、Cr^{6+}、Mn^{6+} 和 Mn^{7+} 其中 Ti(Ⅳ)、V(Ⅳ)和 V(Ⅴ)主要的存在形式是其金属氧基（酰基）离子 TiO^{2+}、VO^{2+} 和 VO$_2^+$；Cr(Ⅵ)、Mn(Ⅵ)和 Mn(Ⅶ)则只形成阴离子盐。Cr(Ⅵ)的阴离子盐在水溶性方面类似于硫酸盐。例如 Ca^{2+}、Sr^{2+}、Ba^{2+}、Pb^{2+} 及 Ag$^+$ 盐均难溶于水，因此可借助于 BaCrO$_4$（柠檬黄色）、PbCrO$_4$（黄色）或 Ag$_2$CrO$_4$（砖红色）沉淀的形成来鉴定 CrO$_4^{2-}$（或 Cr$_2$O$_7^{2-}$）。

此外，周期表中 ⅤB 和 ⅥB 元素的含氧酸常缩合成偏酸和多酸，例如铬酸根 CrO$_4^{2-}$（黄色）在 pH<5 时缩合的主要产物为二铬酸根，即重铬酸根 Cr$_2$O$_7^{2-}$（橙色），在更强的酸性溶液中还可以形成三铬酸根 Cr$_3$O$_{10}^{2-}$、四铬酸根 Cr$_4$O$_{13}^{2-}$ 等。

对于难溶盐来说，多酸盐的溶解度比单酸盐大。因此，加 Ba^{2+}、Pb^{2+} 或 Ag$^+$ 于 CrO$_4^{2-}$ 和 Cr$_2$O$_7^{2-}$ 等的混合溶液中，总是析出难溶的铬酸盐沉淀。

3. 配位化合物

在这些元素的阳离子所形成的配合物中，以 Cr^{3+} 的配合物为多，也较为重要。例如，CrCl$_3 \cdot$ 6H$_2$O 有 3 种颜色的晶体，对应 3 种不同的配位化合物（Cr^{3+} 的配位数为 6）：

　　　　[Cr(H$_2$O)$_6$]Cl$_3$　　　　[Cr(H$_2$O)$_5$Cl]Cl $\cdot$ H$_2$O　　　　[Cr(H$_2$O)$_4$Cl$_2$]Cl $\cdot$ 2H$_2$O
　　　　（蓝紫色）　　　　　　　（蓝绿色）　　　　　　　　　　（灰绿色）

因此，Cr^{3+} 的水溶液往往因其水合程度不同而显不同的颜色。

钛和钒的亲氧能力较强，因此在水溶液中多以较稳定的氧基(酰基)离子或含氧酸根存在，不易形成其他较稳定的配合物，只有 X^- 在较浓时才形成若干卤合离子，例如 $[TiCl_2]^{2+}$、$[TiCl_3]^+$ 或 $[Ti(OH)Cl_5]^{2-}$、$[TiCl_6]^{2-}$ 等。此外在酸性溶液中，Ti(Ⅳ)、V(Ⅳ)、V(Ⅴ) 和 Cr(Ⅵ) 可以与 H_2O_2 形成有特征颜色的配合物：

$$TiO^{2+} + H_2O_2 \rightleftharpoons [TiO(H_2O_2)]^{2+}$$
（橘黄色）

$$VO^{2+} + H_2O_2 \rightleftharpoons [VO(H_2O_2)]^{2+}$$
（红棕色）

$$VO_2^+ + 2H_2O_2 \rightleftharpoons [VO_2(H_2O_2)_2]^+$$
（黄色）

$$Cr_2O_7^{2-} + 4H_2O_2 + 6H^+ \rightleftharpoons 2[CrO_2(H_2O_2)_2]^{2+} + 3H_2O$$

$$Cr_2O_7^{2-} + 4H_2O_2 + 2H^+ \rightleftharpoons 2CrO(O_2)_2 + 5H_2O$$
（蓝色）

向 $[TiO(H_2O_2)]^{2+}$ 溶液中加入氨水，形成过氧钛酸 H_4TiO_5 [即 $TiO(H_2O_2)(OH)_2$] 黄色沉淀，此法鉴定钛很灵敏：

$$[TiO(H_2O_2)]^{2+} + 2NH_3 \cdot H_2O \rightleftharpoons H_4TiO_5 + 2NH_4^+$$

Cr(Ⅵ) 的 H_2O_2 配合物既可鉴定 H_2O_2，也用于鉴定 Cr(Ⅵ)，此配合物易溶于乙醚，在乙醚中较稳定。故鉴定时加乙醚萃取可提高灵敏度。

4. 氧化还原性

Ti、V、Cr、Mn 的标准电极电位 $E_A^{\ominus}$ 图（方括号内为碱性介质的电位，单位：V）如下：

$$TiO^{2+} \xrightarrow{+0.1} TiO^{3+} \xrightarrow{-0.37} Ti^{2+} \xrightarrow{-1.63} Ti$$
（无色）　　　（紫色）　　　（褐色）　　　（银灰色）

$$VO_2^+ \xrightarrow{+1.0} VO^{2+} \xrightarrow{+0.361} V^{3+} \xrightarrow{-0.255} V^{2+} \xrightarrow{-1.18} V$$
（浅黄色）　　（蓝色）　　（绿色）　　（紫色）　　（浅灰色）

$$Cr_2O_7^{2-}(橙红) \xrightarrow[\overline{-0.13}]{+1.33} Cr^{3+}(蓝紫色) \xrightarrow[\overline{0.8}]{-0.41} Cr^{2+}(天蓝) \xrightarrow[\overline{-1.4}]{-0.91} Cr$$
$[Cr_2O_7^{2-}]$(黄色)　　　　$[CrO_2^-]$(绿色)　　　　$[Cr(OH)_2]$

$$MnO_4^-(紫红) \xrightarrow[\overline{+0.564}]{+0.564} MnO_4^{2-}(绿色) \xrightarrow[\overline{+0.60}]{+2.26} MnO_2(棕黑) \xrightarrow[\overline{-0.2}]{+0.95}$$
　　　　　　　　$[MnO_4^{2-}]$(绿色)

$$Mn^{3+}(红色) \xrightarrow[\overline{+0.1}]{+1.51} Mn^{2+}(粉红色) \xrightarrow[\overline{-1.55}]{-1.18} Mn$$
$[Mn(OH)_3]$　　　　$[Mn(OH)_2]$(白色)

在酸性溶液中除 Mn^{2+} 外，M^{2+} 都是相当强的还原剂，都能从水溶液中还原出 H_2：

$$2M^{2+} + 2H^+ \rightleftharpoons 2M^{3+} + H_2\uparrow$$

Cr^{3+} 是比较稳定的，其余 M^{3+} 均不稳定，Ti^{3+} 和 V^{3+} 易被空气中的 O_2 氧化，而 Mn^{3+} 却发生歧化反应：

$$4M^{3+} + O_2 + 2H_2O \rightleftharpoons 4MO^{2+} + 4H^+ \quad (M = Ti, V)$$

$$2Mn^{3+} + 2H_2O = MnO_2\downarrow + Mn^{2+} + 4H^+$$

锰除了 Mn(Ⅲ)容易发生歧化反应外，Mn(Ⅵ)也发生歧化反应。另外，除钛外，它们的高氧化态都是较强的氧化剂，而且氧化性依次增强。

【器材与试剂】

1. 器材：酒精灯，坩埚，坩埚钳，石棉网，三脚架，pH 试纸，淀粉-KI 试纸。
2. 试剂：新鲜配制溶液：Cl_2 水，H_2O_2（质量分数为 3%），Na_2S（0.5 mol·L^{-1}）。
H_2SO_4（3 mol·L^{-1}，浓），HCl（6 mol·L^{-1}，浓），HNO_3（6 mol·L^{-1}，浓），$MnSO_4$（0.1 mol·L^{-1}），$Fe(NO_3)_3$（0.1 mol·L^{-1}），MnO_2(s)，$Cr_2(SO_4)_3$（0.1 mol·L^{-1}），$TiOSO_4$（0.1 mol·L^{-1}），$KMnO_4$（0.01 mol·L^{-1}），Na_2SO_3（0.1 mol·L^{-1}），NaOH（0.1 mol·L^{-1}，6 mol·L^{-1}），$NH_3·H_2O$（6 mol·L^{-1}），KSCN（饱和），K_2CrO_4（0.1 mol·L^{-1}），$K_2Cr_2O_7$（0.1 mol·L^{-1}），$AgNO_3$（0.1 mol·L^{-1}），$Pb(NO_3)_2$（0.1 mol·L^{-1}），$FeSO_4$（0.1 mol·L^{-1}），Fe（粉），NH_4VO_3(s)，Na_2CO_3（0.1 mol·L^{-1}），Zn（粒），$KMnO_4$(s)，$BaCl_2$（0.1 mol·L^{-1}），$CuCl_2$（0.5 mol·L^{-1}）。

【实验内容】

1. Cr 的化合物

（1）$Cr(OH)_3$ 的生成与性质：向 $Cr_2(SO_4)_3$ 溶液中滴加 0.1 mol·L^{-1} NaOH 溶液，观察沉淀的生成。离心分离，沉淀分别做下面实验：

① 与稀硫酸作用；
② 与 6 mol·L^{-1} 的 NaOH 溶液作用；
③ 与 H_2O_2 溶液作用。

（2）Cr^{3+} 的水解与鉴定：检验 $Cr_2(SO_4)_3$ 溶液的 pH，然后分别向盛有少量 $Cr_2(SO_4)_3$ 溶液的试管中滴加 Na_2S 溶液和 Na_2CO_3 溶液，观察现象，通过试验证实所得沉淀均为 $Cr(OH)_3$。

（3）CrO_4^{2-} 与 $Cr_2O_7^{2-}$ 的相互转化：观察并比较 K_2CrO_4 和 $K_2Cr_2O_7$ 溶液的颜色，然后分别向 K_2CrO_4 溶液中滴加 6 mol·L^{-1} 的 HNO_3 溶液，向 $K_2Cr_2O_7$ 溶液中滴加 6 mol·L^{-1} NaOH 溶液，再比较其颜色变化。

（4）难溶性铬酸盐：分别向 K_2CrO_4 和 $K_2Cr_2O_7$ 溶液中滴加 $BaCl_2$ 溶液，观察并比较沉淀的颜色。
用 $AgNO_3$ 和 $Pb(NO_3)_2$ 代替 $BaCl_2$ 分别做同样的试验。

2. Mn 的化合物

（1）Mn(Ⅱ)的化合物及其性质：检验 $MnSO_4$ 溶液的 pH，分别取少量 $MnSO_4$ 溶液做下列试验：

① 与适当的试剂反应获得 $Mn(OH)_2$ 沉淀，观察沉淀在空气中的变化；
② 与 $KMnO_4$ 溶液作用，观察现象；
③ 滴加适量的 Na_2S 溶液，观察沉淀的生成，试设法验证沉淀是 MnS。

（2）Mn(Ⅵ)的生成和性质：取少量 MnO_2 固体加入适量 6 mol·L^{-1} NaOH 溶液，再滴加 $KMnO_4$ 溶液，微热，观察颜色变化，把溶液分成 3 份，分别试验溶液与氯水、Na_2SO_3 溶液和稀硫酸的作用，观察并比较其颜色变化。

（3）Mn(Ⅶ)的氧化性：用 3 支试管，各取少量的 $KMnO_4$ 溶液，其中 1 支用 3 mol·L^{-1} H_2SO_4 酸化，1 支用 6 mol·L^{-1} NaOH 碱化，然后分别滴加 Na_2SO_3 溶液，观察比较 $KMnO_4$ 在不同介质中与 Na_2SO_3 的作用情况。

3. Ti 的化合物

（1）Ti(OH)$_4$的生成和性质：取适量的 TiOSO$_4$溶液，加入 6 mol·L^{-1}氨水，观察沉淀的颜色，将沉淀分成两份，分别试验沉淀在 3 mol·L^{-1} H$_2$SO$_4$和 6 mol·L^{-1} NaOH 溶液中的溶解情况。

（2）TiO^{2+}的氧化还原性质：取适量 TiOSO$_4$溶液，加入少量铁粉，观察紫色的出现。另取一试管，分别滴加 KSCN 溶液和 Fe(NO$_3$)$_3$溶液各 1 滴，把上述紫色溶液滴加到 Fe^{3+}-SCN$^-$溶液中，观察颜色变化。

（3）Ti(Ⅲ)的性质：在 TiOSO$_4$溶液中加 2 颗锌粒，观察溶液颜色的变化。静置 2 分钟后将清液分成两份，分别加 Na$_2$CO$_3$溶液和 CuCl$_2$溶液。

4. V 的化合物

（1）钒(Ⅴ)的氧化性：取饱和 NH$_3$VO$_3$溶液 1～2 mL（自配，用 NH$_3$VO$_3$固体和 6 mol·L^{-1} HCl 溶液配成）。加入一小颗 Zn 粒，放置，并仔细观察溶液颜色的变化。

（2）V$_2$O$_5$的生成和性质：于坩埚中盛少量 NH$_3$VO$_3$固体，小火加热并不断搅拌，根据固体颜色的变化来判断反应的发生和完成。冷却，将固体分成 3 份，一份加浓硫酸，观察固体是否溶解；另一份与 6 mol·L^{-1} NaOH 溶液作用并加热，观察现象；第三份加入浓盐酸并煮沸，观察固体的溶解及颜色变化，怎样证明有氯气放出？

【思考题】

1. 怎样证明 Cr$_2$(SO$_4$)$_3$与 Na$_2$S 溶液作用所生成的沉淀不是硫化物？
2. 怎样配制 NH$_3$VO$_3$饱和溶液？
3. 酸化的 TiOSO$_4$溶液与锌粒作用得到紫色溶液，此紫色溶液在空气中放置一段时间后紫色又褪去，为什么？
4. 根据数据说明在酸性溶液中，处于中间氧化值的钛、钒能否发生歧化反应。
5. 总结实验中有关反应，指出哪些反应可用于鉴定 Ti(Ⅳ)、V(Ⅴ)和 Cr(Ⅵ)，写出反应的化学方程式，注明反应条件，并指出钛、铬、钒在鉴定反应产物中的氧化值。

- 元素毒性
- 化学元素

风险评估及控制工作表

实验任务: 钛、钒、铬和锰的化合物

评估人: 吴泽颖、王红松、吴星星

评估日期: 2021年7月1日

过程步骤	危害鉴别					评估风险	降低或消除风险	安全措施			风险评估				措施	
	化合物名称	类别	浓度	用量	化学品安全技术说明书	可能的伤害种类		技术	防护眼镜	手套	通风橱	可能性	危害	风险	危害级别	紧急情况、泼洒	特殊备注
		或物理危害										1~4	1~4		高中低		

A: Cr 的化合物

① Cr(OH)₃ 的生成与性质: 向 Cr₂(SO₄)₃ 溶液中滴加 0.1 mol·L⁻¹ NaOH 溶液,观察沉淀的生成,离心分离,沉淀分别做下面实验:
① 与稀硫酸作用;
② 与 6 mol·L⁻¹ 的硫酸溶液作用;
③ 与 H₂O₂ 溶液作用。

化合物	类别	浓度	用量	MSDS	伤害	风险	眼镜	手套	橱	可能	危害	风险	级别	措施
Cr₂(SO₄)₃	—	0.1 M	5 mL	√	皮肤灼伤和眼损伤	关注	√	√		1	3	3	低	大量冲洗
NaOH	8	0.1 M	20 mL	√	皮肤灼伤和眼损伤	关注	√	√		1	3	3	低	大量冲洗
稀硫酸	8		少量	√	皮肤灼伤和眼损伤	关注	√	√		1	2	2	低	大量冲洗
NaOH	8	6 M	5 mL	√	皮肤灼伤和眼损伤	关注	√	√		1	4	4	低	大量冲洗
H₂O₂	5.1+8	3%	20 mL	√	皮肤灼伤和眼损伤	关注	√	√		1	3	3	低	大量冲洗

Ⅱ) Cr³⁺ 的水解: 检验 Cr₂(SO₄)₃ 溶液的 pH,然后向试管中滴加 Na₂S 溶液和 Na₂CO₃ 溶液,观察现象,通过试验证实所得沉淀均为 Cr(OH)₃。

| Na₂S | 4.2 | 0.5 M | 5 mL | √ | 皮肤灼伤和眼损伤 | 关注 | √ | √ | | 1 | 3 | 3 | 低 | 大量冲洗 |

Ⅲ) CrO₄²⁻ 与 Cr₂O₇²⁻ 的相互转化: 观察并比较 K₂CrO₄ 和 K₂Cr₂O₇ 溶液的颜色,然后分别向 K₂CrO₄ 溶液中滴加 6 mol·L⁻¹ 的 HNO₃,向 K₂Cr₂O₇ 溶液中滴加 6 mol·L⁻¹ NaOH 溶液,再比较其颜色变化。

K₂CrO₄	5.1+6.1	0.1 M	5 mL	√	皮肤/眼刺激	关注	√	√		1	3	3	低	大量冲洗
K₂Cr₂O₇	9	0.1 M	5 mL	√	皮肤灼伤和眼损伤	关注	√	√		1	2	2	低	大量冲洗
HNO₃	8+5.1	6 M	适量	√	皮肤灼伤和眼损伤	关注	√	√		1	4	4	低	大量冲洗
NaOH	8	6 M	适量	√	皮肤灼伤和眼损伤	关注	√	√		1	3	3	低	大量冲洗

续表

风险评估及控制工作表

实验任务：钛、钒、铬和锰的化合物
评估人：吴泽颖，王红松，吴星星
评估日期：2021年7月1日

过程步骤	化合物名称	危害鉴别		化学品安全技术说明书	健康风险评估		安全措施				评估清单/风险评估	风险评估			危害级别 高/中/低	措施 紧急情况/泼洒	特殊备注
		类别	液度 用量		可能的伤害种类	降低或消除风险	技术	防护眼镜	手套	通风橱		可能性 1~4	危害 1~4	风险			
		物理危害															

IV) 难溶性铬酸盐：分别向 K_2CrO_4 和 $K_2Cr_2O_7$ 溶液中滴加 $BaCl_2$ 溶液，观察并比较沉淀的颜色。用 $AgNO_3$ 和 $Pb(NO_3)_2$ 代替 $BaCl_2$ 别做同样的试验。

	$BaCl_2$	—	0.1 M 适量	√	—	关注		√	√			1	1	1	低	大量冲洗	
	$AgNO_3$	5.1	0.1 M 适量	√	皮肤灼伤和眼损伤	关注		√	√			1	3	3	低	大量冲洗	
	$Pb(NO_3)_2$	5.1+6.1	0.1 M 适量	√	皮肤/眼刺激	关注		√	√			1	1	1	低	大量冲洗	

B：Mn 的化合物

I) Mn(II) 的化合物及其性质：检验 $MnSO_4$ 溶液的 pH，分别取少量 $MnSO_4$ 溶液做下列试验：
① 与适当的试剂反应获得 $Mn(OH)_2$ 沉淀，观察沉淀在空气中的变化；
② 与 $KMnO_4$ 溶液作用，观察现象；
③ 滴加适量的 Na_2S 溶液，观察沉淀的生成，试设法验证沉淀是 MnS。

	$MnSO_4$	—	0.1 M 适量	√	—	关注		√	√			1	1	1	低	大量冲洗	
	$KMnO_4$	5.1	0.01 M 适量	√	皮肤灼伤和眼损伤	关注		√	√			1	1	1	低	大量冲洗	
	Na_2S	4.2	0.5 M 适量	√	皮肤灼伤和眼损伤	关注		√	√			1	1	1	低	大量冲洗	

II) Mn(VI) 的生成和性质：取少量 MnO_2 固体加入适量 6 mol·L^{-1} NaOH 溶液，再滴加 $KMnO_4$ 溶液，微热，观察颜色变化，把溶液分成3份，分别试验溶液与氯水、Na_2SO_3 溶液和稀硫酸的作用，观察并比较其颜色变化

| | MnO_2 固体 | — | 少量 | √ | 吞咽有害 | 关注 | | √ | √ | | | 1 | 1 | 1 | 低 | 立即就医 | |
| | 氯水 | 2.3+5.1+8 | 少量 | √ | 皮肤/眼刺激 | 关注 | | √ | √ | | | 3 | 3 | 3 | 低 | 大量冲洗 | |

风险评估及控制工作表

铁、钒、铬和锰的化合物

评估人：吴泽颖、王红松、吴星星

评估日期：2021年7月1日

化合物名称	危害鉴别			化学品安全技术说明书	健康风险 可能的伤害种类	降低或消除风险	安全措施				评估清单/风险评估 评估日期	风险评估			紧急情况/泼洒特殊备注 措施
	类别	浓度	用量				技术	防护眼镜	手套	通风橱	可能性 1~4	危害 1~4	风险	危害级别 高中低	
物理危害															
H_2SO_4	8	3 M	适量	√	皮肤灼伤和眼损伤	关注		√	√		1	4	4	低	大量冲洗

过程步骤：

Ⅲ) Mn(Ⅶ)的氧化性：用3支试管，各取少量的$KMnO_4$溶液，其中1支用3 mol·L^{-1} H_2SO_4酸化，1支用6 mol·L^{-1} NaOH碱化，然后分别滴加Na_2SO_3溶液，观察比较$KMnO_4$在不同介质中与Na_2SO_3的作用情况。

C: Ti的化合物

化合物名称	类别	浓度	用量	化学品安全技术说明书	可能的伤害种类	降低或消除风险	技术	防护眼镜	手套	通风橱	可能性	危害	风险	危害级别	措施
$TiOSO_4$	—	0.1 M	适量	√	—	关注		√	√		1	2	2	低	大量冲洗
氨水	8	6 M	适量	√	皮肤灼伤和眼损伤	关注		√	√		1	3	3	低	大量冲洗
饱和KSCN溶液	—		1滴	√	眼损伤、皮肤接触有害	关注		√	√		1	1	1	低	大量冲洗
铁粉	—		少量	√	—	关注		√	√		1	1	1	低	大量冲洗
$Fe(NO_3)_3$	5.1	0.1 M	1滴	√	—	关注		√	√		1	1	1	低	大量冲洗

过程步骤：

Ⅰ) Ti(OH)$_4$的生成和性质：取适量的$TiOSO_4$溶液，加入6 mol·L^{-1}氨水，观察沉淀的颜色，将沉淀分成两份，分别试验在3 mol·L^{-1} H_2SO_4和6 mol·L^{-1} NaOH溶液中的溶解情况。

Ⅱ) TiO^{2+}的氧化还原性质：取适量$TiOSO_4$溶液，加入少量铁粉，观察紫色的出现。另取一试管，分别滴加KSCN溶液和$Fe(NO_3)_3$溶液各1滴，把上述紫色溶液滴加到Fe^{3+} – SCN^-溶液中，观察颜色变化。

续 表

钛、钒、铬和锰的化合物

风险评估及控制工作表 评估人：吴泽颖，王红松，吴昊星 评估日期 2021 年 7 月 1 日

实验任务															
过程步骤	危害鉴别				健康风险		降低或消除风险	安全措施				风险评估			措施
	化合物名称	类别	液度	用量	化学品安全技术说明书	可能的伤害种类		技术	防护眼镜	手套	通风橱	清单/风险评估	可能性	危害	风险	危害级别	紧急情况/泼洒 特殊备注

					或物理危害								1~4	1~4	1	高中低	
Ⅲ) Ti(Ⅲ) 的性质：在 TiOSO₄ 溶液中加 2 颗锌粒，观察溶液颜色的变化。静置 2 分钟后将清液分成两份，分别加 Na₂CO₃ 溶液和 CuCl₂ 溶液。	Zn 粒	4.3+4.2		2 颗	√	—	关注		√	√			1	1	1	低	大量冲洗
	Na₂CO₃	—	0.1 M	适量	√	眼刺激	关注		√	√			1	1	1	低	大量冲洗
	CuCl₂	—	0.5 M	适量	√	—	关注		√	√			1	1	1	低	大量冲洗

D: V 的化合物

Ⅰ) 钒（V）的氧化性：取饱和 NH₃VO₃ 溶液 1~2 mL（自配，用 NH₃VO₃ 固体和 6 mol·L⁻¹ HCl 溶液配成。加入一小颗 Zn 粒，放置，并仔细观察溶液颜色的变化。	饱和 NH₃VO₃ 溶液	—		1~2 mL	√	皮肤/眼刺激	关注		√	√	√		1	3	3	低	大量冲洗
	Zn 粒	4.3+4.2		少量	√	—	关注		√	√			1	1	1	低	大量冲洗

续表

风险评估及控制工作表

实验任务	钛、钒、铬和锰的化合物				评估人	吴泽颖,王红松,吴星星				评估日期	2021年7月1日			

过程步骤	危害鉴别				健康风险	降低或消除风险	安全措施			评估清单/风险评估	风险评估			措施
	化合物名称	类别	浓度	用量	化学品安全技术说明书	或物理危害	可能的伤害种类		技术	防护眼镜	手套	通风橱	可能性 1~4	危害 1~4	风险	危害级别 高中低	紧急情况/泼洒	特殊备注

II) V_2O_5 的生成和性质：干坩埚中盛少量 NH_3VO_3 固体，小火加热并不断搅拌，根据固体颜色的变化来判断反应的发生和完成。冷却，将固体分成3份，一份与浓硫酸作用，另一份与 $6\ mol\cdot L^{-1}$ NaOH 溶液作用（加热），第三份加入浓盐酸，检查 Cl_2 颜色变化及 Cl_2 的生成。

	NH_3VO_3 固体	—			√	皮肤/眼刺激	关注			√	√		1	3	3	低	大量冲洗	
	酒精灯					烧伤	关注		√	√			1	3	3	低	立刻向指导老师或准备实验人员报告	
实验产物	Cl_2		2.3+ 5.1+ 8			皮肤/眼刺激	关注			√	√		1	4	4	低	大量冲洗	

实验废弃物	
整个过程的最终评估	低风险（按照正确流程操作）

签名	实验老师：	实验室主任/职业安全健康(OHS)代表：	学生/操作人：

Experiment 9　Compounds of Ti, V, Cr and Mn

【Objectives】

1. To grasp the properties of some important compounds of Ti, V, Cr and Mn (such as hydroxides, salts, coordination compounds).
2. To grasp the oxidation-reduction quality of Ti, V, Cr and Mn.

【Principles】

1. Electron configurations of Ti, V, Cr and Mn in the periodic table

Table 9-1 lists the electron configurations of Ti, V Cr and Mn in the periodic table.

Table 9-1　Electron configurations of Ti, V Cr and Mn in the periodic table

Period	ⅣB	ⅤB	ⅥB	ⅦB
4	22 Ti [Ar]$3d^2 4s^2$	23 V [Ar]$3d^3 4s^2$	24 Cr [Ar]$3d^5 4s^1$	25 Mn [Ar]$3d^5 4s^2$

2. Hydroxides and salts

For Ti, V, Cr and Mn, their hydroxides $M(OH)_2$ are all insoluble and medium strong bases, and their hydroxides $M(OH)_3$ are weak bases. $Cr(OH)_3$ is an amphoteric hydroxide. Cr(Ⅲ) can form both cation salts and anion salts. When their oxidation numbers are high, they bond strongly to oxygen. So Ti^{4+}, V^{4+}, V^{5+}, Cr^{6+}, Mn^{6+} and Mn^{7+} can't be present even in strong acid solutions. Therefore, TiO^{2+}, VO^{2+} and VO^{2+} VO_2^+ are the forms present for Ti(Ⅳ), V(Ⅳ) and V(Ⅴ). Cr(Ⅵ), Mn(Ⅵ) and Mn(Ⅶ) can only form anion salts. The solubilities of anion salts of Cr(Ⅵ) are similar to that of sulfates in water. For example, the sulfates of Ca^{2+}, Sr^{2+}, Ba^{2+}, Pb^{2+} and Ag^+ are insoluble. Therefore, the formation of precipitates of $BaCrO_4$ (lemon yellow), $PbCrO_4$ (yellow) and Ag_2CrO_4 (brick red) can be used to identify CrO_4^{2-} and $Cr_2O_7^{2-}$.

The oxyacids of Group ⅤB and ⅥB usually condense to meta-acids and polyacids. For example, yellow CrO_4^{2-} can turn into orange $Cr_2O_7^{2-}$ when the pH is less than 5. It can turn into $Cr_3O_{10}^{2-}$ and $Cr_4O_{13}^{2-}$ when the pH is lower.

The solubilities of $Cr_2O_7^{2-}$ salts are greater than that of CrO_4^{2-} salts. So precipitates of CrO_4^{2-} salts will form if Ba^{2+}, Pb^{2+} or Ag^+ is added to the mixture of CrO_4^{2-} and $Cr_2O_7^{2-}$.

3. Coordination compounds

The coordination compounds of Cr^{3+} are important. For example, $CrCl_3 \cdot 6H_2O$ represents three complexes, which are three crystals with different colors. The coordination number of Cr^{3+} is 6.

$$[Cr(H_2O)_6]Cl_3 \qquad [Cr(H_2O)_5Cl]Cl_2 \cdot H_2O \qquad [Cr(H_2O)_4Cl_2]Cl \cdot 2H_2O$$
$$\text{(purple blue)} \qquad \text{(blue green)} \qquad \text{(greyish green)}$$

So, the aqueous solutions of Cr^{3+} show different colors.

Ti and V bond strongly to oxygen, so TiO^{2+}, VO^{2+} and VO_2^+ are stable in aqueous solutions.

It is not easy for Ti and V to form stable coordination compounds. But they can form some complex ions with halogen ions which have high concentrations, such as $[TiCl_2]^{2+}$, $[TiCl_3]^+$, $[Ti(OH)Cl_5]^{2-}$, $[TiCl_6]^{2-}$, $[TiF_6]^{2-}$, etc. In acidic solution, Ti(Ⅳ), V(Ⅳ), V(Ⅴ) and Cr(Ⅵ) can react with H_2O_2 to form complex ions with typical colors.

$$TiO^{2+} + H_2O_2 \rightleftharpoons [TiO(H_2O_2)]^{2+}$$
$$\text{(orange)}$$

$$VO^{2+} + H_2O_2 \rightleftharpoons [VO(H_2O_2)]^{2+}$$
$$\text{(red brown)}$$

$$VO_2^+ + 2H_2O_2 \rightleftharpoons [VO_2(H_2O_2)_2]^+$$
$$\text{(yellow)}$$

$$Cr_2O_7^{2-} + 4H_2O_2 + 6H^+ \rightleftharpoons 2[CrO_2(H_2O_2)_2]^{2+} + 3H_2O$$

$$Cr_2O_7^{2-} + 4H_2O_2 + 2H^+ \rightleftharpoons 2CrO(O_2)_2 + 5H_2O$$
$$\text{(blue)}$$

Add ammonia water to $[TiO(H_2O_2)]^{2+}$ solution, and yellow H_4TiO_5 [or $TiO(H_2O_2)(OH)_2$] will be formed. This reaction is used to identify Ti.

$$[TiO(H_2O_2)]^{2+} + 2NH_3 \cdot H_2O \rightleftharpoons H_4TiO_5 + 2NH_4^+$$

Cr(Ⅵ) reacts with H_2O_2 to form a complex which can be used to identify both H_2O_2 and Cr(Ⅵ). This complex dissolves and is stable in diethyl ether. So diethyl ether is used to improve the sensitivity.

4. Oxidation-reduction properties

The standard electrode potentials ($E_A^\ominus/V$) of Ti、V、Cr、Mn are as follows (data or compounds in syuare brackets are the standard potentials $E_B^\ominus/V$) or compounds in basic solution):

$$TiO^{2+} \xrightarrow{+0.1} TiO^{3+} \xrightarrow{-0.37} Ti^{2+} \xrightarrow{-1.63} Ti$$
$$\text{(colorless)} \quad \text{(purple)} \quad \text{(brown)} \quad \text{(silver grey)}$$

$$VO_2^+ \xrightarrow{+1.0} VO^{2+} \xrightarrow{+0.361} V^{3+} \xrightarrow{-0.255} V^{2+} \xrightarrow{-1.18} V$$
$$\text{(light yellow)} \quad \text{(blue)} \quad \text{(green)} \quad \text{(purple)} \quad \text{(light grey)}$$

$$Cr_2O_7^{2-}\text{(orange)} \xrightarrow{+1.33} Cr^{3+}\text{(purple blue)} \xrightarrow{-0.41} Cr^{2+}\text{(sky blue)} \xrightarrow{-0.91} Cr$$
$$[Cr_2O_7^{2-}]\text{(yellow)} \xrightarrow{[-0.13]} [CrO_2^-]\text{(green)} \xrightarrow{[0.8]} [Cr(OH)_2] \xrightarrow{[-1.4]}$$

$$MnO_4^-\text{(purple)} \xrightarrow{+0.564} MnO_4^{2-}\text{(green)} \xrightarrow{+2.26} MnO_2\text{(brown black)} \xrightarrow{+0.95}$$
$$\xrightarrow{[+0.564]} [MnO_4^{2-}]\text{(green)} \xrightarrow{[+0.60]} [MnO_2] \xrightarrow{[-0.2]}$$

$$Mn^{3+}\text{(red)} \xrightarrow{+1.51} Mn^{2+}\text{(pink)} \xrightarrow{-1.18} Mn$$
$$[Mn(OH)_3] \xrightarrow{[+0.1]} [Mn(OH)_2]\text{(white)} \xrightarrow{[-1.55]}$$

In acidic solution, all M^{2+} ions are strong reducing agents except Mn^{2+}.

$$2M^{2+} + 2H^+ = 2M^{3+} + H_2 \uparrow$$

Cr^{3+} is stable, but other M^{3+} ions are not. Ti^{3+} and V^{3+} are easily oxidized by O_2 in the air, but Mn^{3+} disproportionates:

$$4M^{3+} + O_2 + 2H_2O = 4MO^{2+} + 4H^+ \quad (M = Ti, V)$$

$$2Mn^{3+} + 2H_2O = MnO_2 \downarrow + Mn^{2+} + 4H^+$$

【Apparatus and Chemicals】

1. Apparatus

Alcohol burner; crucible; crucible tongs; gauze with asbestos; tripod; pH test paper; starch test-KI paper.

2. Chemicals

Freshly prepared solutions: chlorine water, H_2O_2(3%, wt/wt), Na_2S (0.5 mol·L^{-1}).

H_2SO_4 (3 mol·L^{-1}, concentrated); HCl (concentrated); HNO_3 (6 mol·L^{-1}, concentrated); $MnSO_4$ (0.1 mol·L^{-1}); $Fe(NO_3)_3$ (0.1 mol·L^{-1}); MnO_2(s); $Cr_2(SO_4)_3$ (0.1 mol·L^{-1}); $TiOSO_4$ (0.1 mol·L^{-1}); $KMnO_4$ (0.01 mol·L^{-1}); Na_2SO_3 (0.1 mol·L^{-1}); NaOH (0.1 mol·L^{-1}, 6 mol·L^{-1}); $NH_3 \cdot H_2O$ (6 mol·L^{-1}); KSCN (saturated); K_2CrO_4 (0.1 mol·L^{-1}); $K_2Cr_2O_7$ (0.1 mol·L^{-1}); $AgNO_3$ (0.1 mol·L^{-1}); $Pb(NO_3)_2$ (0.1 mol·L^{-1}); $FeSO_4$ (0.1 mol·L^{-1}); Fe powder; NH_4VO_3(s); Na_2CO_3 (0.1 mol·L^{-1}); Zn, $KMnO_4$(s); $BaCl_2$ (0.1 mol·L^{-1}); $CuCl_2$ (0.5 mol·L^{-1}).

【Experimental Procedures】

1. Compounds of Cr

(1) The formation and properties of $Cr(OH)_3$: Add 0.1 mol·L^{-1} NaOH solution to $Cr_2(SO_4)_3$ solution. Observe the formation of the precipitate. Centrifuge and separate the precipitate. Use the precipitate to do the following experiments:

① React with dilute H_2SO_4.

② React with 6 mol·L^{-1} NaOH.

③ React with H_2O_2 solution.

(2) Hydrolysis and appraisal of Cr^{3+}: Test the pH of $Cr_2(SO_4)_3$ solution. Put $Cr_2(SO_4)_3$ solution in two test tubes. Then add Na_2S solution to one test tube, and Na_2CO_3 solution to the other. Prove the precipitates are $Cr(OH)_3$.

(3) Transformation between CrO_4^{2-} and $Cr_2O_7^{2-}$: Observe and compare the colors of K_2CrO_4 solution and $K_2Cr_2O_7$ solution. Then add 6 mol·L^{-1} HNO_3 to K_2CrO_4 solution, adding 6 mol·L^{-1} NaOH to $K_2Cr_2O_7$ solution. Observe and compare the color changes.

(4) Insoluble chromate: Add $BaCl_2$ solution to K_2CrO_4 and $K_2Cr_2O_7$ respectively, and compare the colors of the precipitates. Use $AgNO_3$ and $Pb(NO_3)_2$ to replace $BaCl_2$ solution to replicate the above experiments.

(5) Oxidizability of Cr(Ⅵ). Add $FeSO_4$ solution to K_2CrO_4 and $K_2Cr_2O_7$ respectively, and add

dilute H_2SO_4. Observe and record your observation.

2. Compounds of Mn

(1) Compounds of Mn(Ⅱ): Test the pH of $MnSO_4$ solution, and then do the following experiments.

① Use a small amount of $MnSO_4$ solution to react with the appropriate agent to produce $Mn(OH)_2$ precipitate. Observe the changes of the precipitate in the air.

② Use a small amount of $MnSO_4$ solution to react with $KMnO_4$ solution. Observe and record your observations.

③ Add an appropriate amount of Na_2S solution to $MnSO_4$ solution. Observe the precipitate, and test if the precipitate is MnS.

(2) Properties of Mn(Ⅵ): Use a small amount of solid MnO_2 to react with an appropriate amount of 6 mol·L^{-1} NaOH, then add $KMnO_4$ to the solution. Slightly heat and observe the color changes. Test the reaction of the solution with ammonia water, Na_2SO_3 solution and dilute H_2SO_4 respectively. Compare the color changes.

(3) Oxidizability of Mn(Ⅶ): Place a small amount of $KMnO_4$ solution into three test tubes. Acidify the first solution with 3 mol·L^{-1} H_2SO_4, and add 6 mol·L^{-1} NaOH to the second one. Then add Na_2SO_3 solution to each test tube. Observe and compare the reaction of $KMnO_4$ in the different media.

3. Compounds of Ti

(1) $Ti(OH)_4$: Add 6 mol·L^{-1} ammonia water to an appropriate amount of $TiOSO_4$ solution. Observe the color of the precipitate. Test the dissolution of the precipitate in 3 mol·L^{-1} H_2SO_4 and 6 mol·L^{-1} NaOH respectively.

(2) Oxidation and reduction of TiO^{2+}: Add Fe powder to an appropriate amount of $TiOSO_4$ solution. Observe the purple color. Put one drop of KSCN solution and one drop of $Fe(NO_3)_3$ solution to a test tube, and add the above purple solution to $Fe^{3+}-SCN^-$ solution. Observe the color change.

(3) The properties of Ti(Ⅲ): Add two granulated zincs into $TiOSO_4$ solution, and then observe. Rest for 2 minutes, divide the liquid into two parts and add Na_2CO_3 and $CuCl_2$, respectively.

4. Compounds of V

(1) Oxidizability of V: In a test tube, transfer 1 to 2 mL of saturated NH_3VO_3 solution which is prepared by using solid NH_3VO_3 and 6 mol·L^{-1} HCl solution. Add a small grain of Zn to the prepared solution and observe the color change.

(2) V_2O_5: Put a small amount of solid NH_3VO_3 in a clean crucible. Heat and stir constantly. Determine the completion of the reaction according to the color change of the solid. Allow the solid to cool. Divide the solid into three portions. React the first portion with concentrated H_2SO_4, the second with 6 mol·L^{-1} NaOH solution while heating, and the third with concentrated HCl. Observe the dissolution of the solid and the color changes of the solutions. Test if Cl_2 is produced.

【Questions】

1. How do we prove the precipitate produced by the reaction of $Cr_2(SO_4)_3$ and Na_2S is not sulfide?

2. How do we prepare saturated NH_3VO_3 solution?

3. Acidified $TiOSO_4$ solution reacts with Zn to produce a purple solution. The purple color fades after it is in the air for a while. Why?

4. According to the data of $E^\ominus$, can the disproportionation reactions of Ti and V occur in an acidic solution?

5. What reactions can be used to identify Ti(IV), V(V) and Cr(VI)? Record the chemical equations. Point out the oxidation number of Ti, V and Cr in the products.

Risk Assessment and Control Worksheet

TASK	Compounds of Ti, V, Cr and Mn					Assessor	Zeying WU, Hongsong WANG, Xingxing WU					Date	1 JUL. 2021				
Process Steps	Hazard Identification					Health Risk	Safety Measures					Checklist～Risk Assessment	Risk Assessment				Action
	Chemical Name or Physical Hazard	Class	Concn.	Amount	MSDS	Type of Injury Possible	To reduce or eliminate the risk	Technique	Glasses	Gloves	Fume Cbd.		Likely 1～4	Hazard 1～4	Risk Score	Risk Level H M L	In case of Emergency/Spill Special Comments
A: Compounds of Cr																	
Ⅰ) $Cr(OH)_3$ Add 0.1 mol・L^{-1} NaOH solution to $Cr_2(SO_4)_3$ solution. Observe the formation of the precipitate. Centrifuge and separate the precipitate. Use the precipitate to do the following experiments: ① React with diluted H_2SO_4. ② React with 6 mol・L^{-1} NaOH. ③ React with H_2O_2 solution.	$Cr_2(SO_4)_3$	—	0.1 M	5 mL	√	Skin burns and eye injuries	Handle with care		√	√			1	3	3	L	Sponge & water
	NaOH	8	0.1 M	20 mL	√	Skin burns and eye injuries	Handle with care		√	√			1	3	3	L	Sponge & water
	Diluted H_2SO_4	8		a little	√	Skin burns and eye injuries	Handle with care		√	√			1	2	2	L	Sponge & water
	NaOH	8	6 M	5 mL	√	Skin burns and eye injuries	Handle with care		√	√			1	4	4	L	Sponge & water
	H_2O_2	5.1 +8	3%	20 mL	√	Skin burns and eye injuries	Handle with care		√	√			1	3	3	L	Sponge & water
Ⅱ) Hydrolysis of Cr^{3+} Test the pH of $Cr_2(SO_4)_3$ solution. Put $Cr_2(SO_4)_3$ solution in two test tubes. Then add Na_2S solution to one test tube, and Na_2CO_3 solution to the other. Prove the precipitates are $Cr(OH)_3$.	Na_2S	4.2	0.5 M	5 mL	√	Skin burns and eye injuries	Handle with care		√	√			1	3	3	L	Sponge & water
Ⅲ) Transformation between CrO_4^{2-} and $Cr_2O_7^{2-}$, Observe and compare the colors of K_2CrO_4 solution and $K_2Cr_2O_7$ solution. Then add 6 mol・L^{-1} HNO_3 to K_2CrO_4 solution, adding 6 mol・L^{-1} NaOH to $K_2Cr_2O_7$ solution. Observe and compare the color changes.	K_2CrO_4	5.1+ 6.1	0.1 M	5 mL	√	Skin and eye irritant	Handle with care		√	√			1	3	3	L	Sponge & water
	$K_2Cr_2O_7$	9	0.1 M	5 mL	√	Skin burns and eye injuries	Handle with care		√	√			1	2	2	L	Sponge & water
	HNO_3	8+ 5.1	6 M	appropriate amount	√	Skin burns and eye injuries	Handle with care		√	√			1	4	4	L	Sponge & water
	NaOH	8	6 M	appropriate amount	√	Skin burns and eye injuries	Handle with care		√	√			1	3	3	L	Sponge & water

Risk Assessment and Control Worksheet

TASK: Compounds of Ti, V, Cr and Mn
Assessor: Zeying WU, Hongsong WANG, Xingxing WU
Date: 1 JUL. 2021

Process Steps	Chemical Name	Class	Concn.	Amount	MSDS	Type of Injury Possible	To reduce or eliminate the risk	Technique	Glasses	Gloves	Fume Cbd.	Likely (1~4)	Hazard (1~4)	Risk Score	Risk Level (H M L)	In case of Emergency/Spill	Special Comments
Ⅳ) Insoluble chromate. Add BaCl₂ solution to K₂CrO₄ and K₂Cr₂O₇ respectively, and compare the colors of the precipitates. Use AgNO₃ and Pb(NO₃)₂ to replace BaCl₂ solution to replicate the above experiments.	BaCl₂	—	0.1 M	appropriate amount	√	—	Handle with care			√		1	1	1	L	Sponge & water	
	AgNO₃	5.1	0.1 M	appropriate amount	√	Skin burns and eye injuries	Handle with care		√	√		1	3	3	L	Sponge & water	
	Pb(NO₃)₂	5.1+6.1	0.1 M	appropriate amount	√	Skin and eye irritant	Handle with care		√	√		1	1	1	L	Sponge & water	
B: Compounds of Mn																	
Ⅰ) Compounds of Mn(Ⅱ). Test the pH of MnSO₄ solution, and then do the following experiments. ① Use a small amount of MnSO₄ solution to react with the appropriate agent to produce Mn(OH)₂ precipitate. Observe the changes of the precipitate in the air. ② Use a small amount of MnSO₄ solution to react with KMnO₄ solution. Observe and record your observations. ③ Add an appropriate amount of Na₂S solution to MnSO₄ solution. Observe the precipitate, and test if the precipitate is MnS.	MnSO₄	—	0.1 M	appropriate amount	√	—	Handle with care		√	√		1	1	1	L	Sponge & water	
	KMnO₄	5.1	0.01 M	appropriate amount	√	—	Handle with care		√	√		1	1	1	L	Sponge & water	
	Na₂S	4.2	0.5 M	appropriate amount	√	Skin burns and eye injuries	Handle with care		√	√		1	1	1	L	Sponge & water	

continued

Risk Assessment and Control Worksheet

TASK	Compounds of Ti, V, Cr and Mn					Assessor	Zeying WU, Hongsong WANG, Xingxing WU					Date	1 JUL. 2021			
Process Steps	Hazard Identification					Health Risk		Safety Measures				Checklist＼Risk Assessment				Action
	Chemical Name	Class or Physical Hazard	Concn.	Amount	MSDS	Type of Injury Possible	To reduce or eliminate the risk	Technique	Glasses	Gloves	Fume Cbd.	Likely 1~4	Hazard 1~4	Risk Score	Risk Level H M L	In case of Emergency/Spill Special Comments
II) Properties of Mn(VI). Use a small amount of solid MnO_2 to react with an appropriate amount of $6\ mol·L^{-1}$ NaOH, then add $KMnO_4$ to the solution. Slightly heat and observe the color changes.	MnO_2	—	—	a little	√	Swallowing harmful	Handle with care		√	√		1	1	1	L	Go to a doctor immediately
Test the reaction of the solution with ammonia, Na_2SO_3 solution and dilute H_2SO_4 respectively. Compare the color changes.	Ammonia	2.3+ 5.1+ 8		a little	√	skin and eye irritant	Handle with care		√	√		1	3	3	L	Sponge & water
III) Oxidizability of Mn(VII). Place a small amount of $KMnO_4$ solution into three test tubes. Acidify the first solution with $3\ mol·L^{-1}\ H_2SO_4$, and add $6\ mol·L^{-1}$ NaOH to the second one. Then add Na_2SO_3 solution to each test tube. Observe and compare the reaction of $KMnO_4$ in the different media.	H_2SO_4	8	3 M	appropriate amount	√	Skin burns and eye injuries	Handle with care		√	√		1	4	4	L	Sponge & water
C: Compounds of Ti																
I) $Ti(OH)_4$. Add $6\ mol·L^{-1}$ ammonia to an appropriate amount of $TiOSO_4$ solution. Observe the color of the precipitate. Test the dissolution of the precipitate in $3\ mol·L^{-1}\ H_2SO_4$ and $6\ mol·L^{-1}$ NaOH respectively.	$TiOSO_4$	—	0.1 M	appropriate amount	√	—	Handle with care		√	√	√	1	2	2	L	Sponge & water
	Ammonia	8	6 M	appropriate amount	√	Skin burns and eye injuries	Handle with care		√	√	√	1	3	3	L	Sponge & water

Risk Assessment and Control Worksheet continued

TASK: Compounds of Ti, V, Cr and Mn **Assessor:** Zeying WU, Hongsong WANG, Xingxing WU **Date:** 1 JUL. 2021

Process Steps	Chemical Name	Hazard Identification — Class	Concn.	Amount	MSDS	Health Risk — Type of Injury Possible	Safety Measures — To reduce or eliminate the risk	Technique	Glasses	Gloves	Fume Cbd.	Checklist\Risk Assessment — Likely (1~4)	Hazard (1~4)	Risk Score	Risk Level (H M L)	In case of Emergency/Spill Action	Special Comments
II) Oxidation and reduction of TiO^{2+}. Add Fe powder to an appropriate amount of $TiOSO_4$ solution. Observe the purple color.	Saturated KSCN solution	—		1 drop	✓	Eye injuries & Harmful by skin contact	Handle with care		✓	✓	✓	1	1	1	L	Sponge & water	
Put one drop of KSCN solution and one drop of $Fe(NO_3)_3$ solution to a test tube, and add the above purple solution to $Fe^{3+} - SCN^-$ solution. Observe the color change.	Iron powder	—		a little	✓	—	Handle with care		✓	✓		1	1	1	L	Sponge & water	
	$Fe(NO_3)_3$	5.1	0.1 M	1 drop	✓	—	Handle with care		✓	✓		1	1	1	L	Sponge & water	
III) The properties of Ti(III). Add two granulated zincs into $TiOSO_4$ solution, and then observe. Rest for 2 minutes, divide the liquid into two parts and add Na_2CO_3 and $CuCl_2$, respectively.	Zn	4.3+ 4.2		2 grains	✓	—	Handle with care		✓	✓		1	1	1	L	Sponge & water	
	Na_2CO_3	—	0.1 M	appropriate amount	✓	Eye irritant	Handle with care		✓	✓		1	1	1	L	Sponge & water	
	$CuCl_2$	—	0.5 M	appropriate amount	✓	—	Handle with care		✓	✓		1	1	1	L	Sponge & water	
D: Compounds of V																	
I) Oxidizability of V. In a test tube, transfer 1 to 2 mL of saturated NH_3VO_3 solution which is prepared by using solid NH_3VO_3 and 6 mol·L^{-1} HCl solution. Add a small grain of Zn to the prepared solution and observe the color change.	Saturated NH_3VO_3 solution	—		1~2 mL	✓	Skin and eye irritant	Handle with care		✓	✓	✓	1	3	3	L	Sponge & water	
	Zn	4.3+ 4.2		a little	✓	—	Handle with care		✓	✓		1	1	1	L	Sponge & water	

continued

Risk Assessment and Control Worksheet

TASK	Compounds of Ti, V, Cr and Mn					Assessor	Zeying WU, Hongsong WANG, Xingxing WU				Date	1 JUL. 2021				
Process Steps	Hazard Identification					Health Risk		Safety Measures				Checklist~Risk Assessment				Action
	Chemical Name	Class	Concn.	Amount	MSDS	Type of Injury Possible	To reduce or eliminate the risk	Technique	Glasses	Gloves	Fume Cbd.		Risk Assessment			In case of Emergency/Spill
			or Physical Hazard									Likely	Hazard	Risk Score	Risk Level	Special Comments
												1~4	1~4		H M L	
II) V_2O_5. Put a small amount of solid NH_3VO_3 in a clean crucible. Heat and stir constantly. Determine the completion of the reaction according to the color change of the solid. Allow the solid to cool. Divde the solid into three portions. React the first portion with concentrated H_2SO_4, the second with 6 mol·L^{-1} NaOH solution while heating, and the third with concentrated HCl. Observe the dissolution of the solid and the color changes of the solutions. Test if Cl_2 is produced.	NH_3VO_3	—			√	Skin and eye irritant	Handle with care		√	√		1	3	3	L	Sponge & water
	Alcohol burner					Burns	Handle with care	√				1	3	3	L	Report immediately to Demonstrator or Prep Room
Experimental Product (s)	Cl_2	2.3+ 5.1+ 8			√	Skin and eye irritant	Handle with care		√	√		1	4	4	L	Sponge & water
Experimental Waste (s)																
Final assessment of overall process	Low risk when correct procedures followed															
Signatures	Supervisor:						Lab Manager/OHS Representative:							Student/Operator:		

实验 10 铁、钴、镍的化合物

【实验目的】

1. 掌握铁、钴、镍氢氧化物的生成和性质。
2. 掌握铁、钴、镍硫化物的生成和性质。
3. 掌握铁、钴、镍配合物的生成和性质。

【实验原理】

铁组元素属于Ⅷ族,包括铁、钴和镍 3 种元素。它们是同一周期的元素,原子结构相似($[Ar]3d^{6\sim 8}4s^2$),原子半径相近(115~117 pm)。

铁组元素常见的盐类是 Fe(Ⅲ)、Fe(Ⅱ)、Co(Ⅱ)和 Ni(Ⅱ)盐,其中硝酸盐、硫酸盐和卤化物均易溶于水。这些阳离子水合时,通常发生颜色变化。例如 Fe^{2+} 由白色变成浅绿色,Co^{2+} 由蓝色变成粉红色,Ni^{2+} 由黄色变成亮绿色。它们的盐从水中结晶析出时,常形成带结晶水的晶体。随着结晶水量的变化,其颜色也发生变化。例如:

$$CoCl_2 \cdot 6H_2O \xrightleftharpoons{520\ ℃} CoCl_2 \cdot 2H_2O \xrightleftharpoons{90\ ℃} CoCl_2 \cdot H_2O \xrightleftharpoons{120\ ℃} CoCl_2$$
（粉红色）　　　　（紫红色）　　　　（蓝紫色）　　　　（蓝色）

$CoCl_2$ 的这一性质常用于实验室显示干燥剂(硅胶)吸水的情况。

Fe^{3+} 盐与 Cr^{3+} 盐很类似。例如,MCl_3 均可双聚形成共边双四面体分子 M_2Cl_6,且易升华;其硫酸盐与碱金属元素的硫酸盐类均能形成 $M_2SO_4 \cdot M_2(SO_4)_3 \cdot 24H_2O$ 的"矾",如与通常净水用的明矾 $K_2SO_4 \cdot Al_2(SO_4)_3 \cdot 24H_2O$ 类似的有钾铁矾 $K_2SO_4 \cdot Fe_2(SO_4)_3 \cdot 24H_2O$ 等。

铁组元素常见的重要难溶盐见表 10-1。

表 10-1 铁组元素常见的难溶盐

阳离子	阴离子			
	S^{2-}	CO_3^{2-}	CrO_4^{2-}	$[Fe(CN)_6]^{4-}$
Fe^{3+}	黑色	—	—	蓝色
Fe^{2+}	黑色	白色	白色	白色
Co^{2+}	黑色	粉红色	浅粉红色	绿色
Ni^{2+}	黑色	浅绿色	浅绿色	浅绿色

铁组元素的阳离子是形成配合物的较好形成体,能形成很多的配合物。唯 Fe^{3+}、Fe^{2+} 与 OH^- 结合的能力较强,它们难以在氨水中形成稳定的氨合离子。常见的稳定配合物见表 10-2。由于配位后发生溶解度和颜色等的改变,常用于离子的分离和鉴定。

表 10-2 铁组元素常见的配合物

中心离子	配体				
	H_2O	CN^-	NH_3	SCN^-	F^-
Fe^{3+}	浅紫色	浅黄色	—	血红色	无色
Fe^{2+}	浅绿色	黄色	—	无色	—
Co^{3+}	—	—	红色	—	—
Co^{2+}	粉红色	紫色	橙色	蓝色	—
Ni^{2+}	绿色	黄色	深蓝色	—	—

铁组元素氢氧化物的酸碱性完全符合一般规律。

M(Ⅵ)的氢氧化物为强酸,故无论在酸性还是碱性溶液中均以酸根 MO_4^{2-} 的形式存在。其中较为重要的是高铁酸盐。FeO_4^{2-} 与 MnO_4^- 一样呈紫红色,与 MnO_4^{2-}、CrO_4^{2-} 和 SO_4^{2-} 等类似,能使 Ba^{2+} 沉淀。

铁组元素的 $M(OH)_3$ 和 $M(OH)_2$ 均为碱,但碱性不强。新沉淀出来的 $Fe(OH)_3$ 还有微弱的酸性,可溶于热的浓 KOH 溶液中,形成铁酸钾 $KFeO_2$。因此铁组元素的可溶性盐均发生水解,且 M^{3+} 盐比 M^{2+} 盐的水解度要大些,尤以 Fe^{3+} 盐突出。Fe^{3+} 盐的水解产物常使其溶液变成棕黄色或棕红色:

$$Fe^{3+} + nH_2O \rightleftharpoons [Fe(OH)_n]^{(3-n)} + nH^+ \quad (n=1, 2, 3)$$

Fe^{3+} 仅存在于强酸性溶液(pH<2)中。稀释或提高溶液的 pH,会析出胶状的红棕色 $Fe(OH)_3$ 沉淀。

铁组元素的电位图 $E_A^\ominus/V$(方括号内注明的是碱性溶液中的状态及 $E_B^\ominus/V$)为:

$$FeO_4^{2-} \xrightarrow[\text{[+0.9]}]{+1.9} \underset{[Fe(OH)_3]}{Fe^{3+}} \xrightarrow[\text{[-0.56]}]{+0.771} \underset{[Fe(OH)_2]}{Fe^{2+}} \xrightarrow[\text{[-0.887]}]{-0.440} Fe$$

$$CoO_2 \xrightarrow[\text{[+0.7]}]{>+1.8} \underset{[Co(OH)_3]}{Co^{3+}} \xrightarrow[\text{[+0.17]}]{+1.82} \underset{[Co(OH)_2]}{Co^{2+}} \xrightarrow[\text{[-0.73]}]{-0.277} Co$$

$$NiO_4^{2-} \xrightarrow[\text{[>+0.4]}]{+1.8} NiO_2 \xrightarrow[\text{[+0.49]}]{+1.68} \underset{[Ni(OH)_2]}{Ni^{2+}} \xrightarrow[\text{[-0.72]}]{-0.250} Ni$$

氧化值大于或等于+4 的铁组元素化合物是很强的氧化剂。在酸性条件下它们的氧化性比 MnO_4^- 还强;在碱性介质中,它们的氧化性大为降低;在强碱中用强氧化剂(NaClO 等)就可以将低氧化值的铁组元素氧化。例如:

$$2Fe(OH)_3 + 3ClO^- + 4OH^- \rightleftharpoons 2FeO_4^{2-} + 3Cl^- + 5H_2O$$

$$Fe_2O_3 + 3KNO_3 + 4KOH \xrightarrow{\text{熔融}} 2K_2FeO_4 + 3KNO_2 + 2H_2O$$

M(Ⅲ)在酸性介质中也是氧化剂,其中 Fe^{3+} 为中强氧化剂;Co^{3+} 为强氧化剂,能氧化水和盐酸;Ni^{3+} 的氧化性更强。在酸性介质中,Fe^{3+} 能被 Fe、Cu、Sn^{2+}、S^{2-}、I^- 或 SO_2 等还原。

若有配体或沉淀剂存在,M(Ⅲ)被配位或形成难溶盐而稳定存在。例如(单位:V):

$$Fe^{3+} \xrightarrow{+0.771} Fe^{2+} \qquad\qquad Co^{3+} \xrightarrow{+1.842} Co^{2+}$$

$$[FeF_6]^{3-} \xrightarrow{-0.135} Fe^{2+} \qquad\qquad [Co(NH_3)_6]^{3+} \xrightarrow{+0.1} [Co(NH_3)_6]^{2+}$$

$$[Fe(CN)_6]^{3-} \xrightarrow{+0.36} [Fe(CN)_6]^{4-} \qquad\qquad [Co(CN)_6]^{3-} \xrightarrow{+0.81} [Co(CN)_6]^{4-}$$

综上所述,铁组元素 M(Ⅲ)、M(Ⅱ)和 M(0)氧化还原性及相应氢氧化物的酸碱性变化规律为:

```
                    在酸性溶液中还原性增强
              ┌─────────────────────────────┐
              │    Fe        Co        Ni   │
              │  (银白色)   (银白色)   (银白色) │
              │                             │
  还原性增强   │   Fe²⁺       Co²⁺      Ni²⁺  │  氧化性增强
              │  (浅绿色)   (粉红色)   (亮绿色) │
              │                             │
              │   Fe³⁺       Co³⁺      Ni³⁺  │
              │  (浅紫色)    (绿色)    (粉红色) │
              └─────────────────────────────┘
                        氧化性增强

                    在碱性溶液中还原性增强
              ┌─────────────────────────────┐
              │    Fe        Co        Ni   │
              │  (银白色)   (银白色)   (银白色) │
              │                             │
还原性增强 碱性增强│ Fe(OH)₂   Co(OH)₂    Ni(OH)₂ │ 氧化性增强 碱性减弱
              │   (白色)    (粉红色)   (浅绿色) │
              │                             │
              │ Fe(OH)₃    Co(OH)₃    Ni(OH)₃│
              │  (红棕色)   (暗棕色)   (灰黑色) │
              └─────────────────────────────┘
                        氧化性增强
```

【铁组离子鉴定】

(1) Fe^{3+}　在弱酸性介质中 Fe^{3+} 与 KSCN 作用形成血红色配离子 $[Fe(SCN)_n]^{(3-n)}$。

$$Fe^{3+} + nKSCN = [Fe(SCN)_n](3-n) + nK^+$$

此法无离子干扰。若其他重金属离子浓度不太高，也可借助 Fe^{3+} 与黄血盐 $K_4[Fe(CN)_6]$ 作用生成深蓝色的普鲁士蓝 $[KFe(CN)_6Fe]_x$ 沉淀来鉴定：

$$xFe^{3+} + xK^+ + x[Fe(CN)_6]^{4-} = [KFe(CN)_6Fe]_x \downarrow$$

(2) Fe^{2+}　Fe^{2+} 与赤血盐 $K_3[Fe(CN)_6]$ 作用生成深蓝色的藤氏蓝 $[KFe(CN)_6Fe]_x$ 沉淀。

$$xFe^{2+} + xK^+ + x[Fe(CN)_6]^{3-} = [KFe(CN)_6Fe]_x \downarrow$$

当重金属离子含量不高时，此法的灵敏度和选择性均较高。

(3) Co^{2+}　Co^{2+} 与 KSCN 作用生成蓝色配离子 $[Co(SCN)_4]^{2-}$。

$$Co^{2+} + 4SCN^- = [Co(SCN)_4]^{2-}$$

该配离子在丙酮中较稳定，故鉴定时需加入丙酮或戊醇。若有 Fe^{3+}，可加入 NaF 以形成无色的 $[FeF_6]^{3-}$ 掩蔽之。

(4) Ni^{2+}　Ni^{2+} 在弱碱性($NH_3 \cdot H_2O$)介质中与丁二酮肟生成鲜红色螯合物沉淀，是鉴定 Ni^{2+} 的灵敏反应。

```
                CH₃—C—NOH              CH₃—C=N    N=C—CH₃
                    |                       |    \ /    |
       Ni²⁺ + 2     |         + 2NH₃ =     |    Ni     |      + 2NH₄⁺
                CH₃—C—NOH              CH₃—C=N    N=C—CH₃
                                           |         |
                                          O—H···H—O
```

【器材与试剂】

1. 器材：KI-淀粉试纸。
2. 试剂：HCl（浓），H_2S（饱和，新配），NaOH（s，6 mol·L^{-1}，新配），H_2SO_4（稀），氨水（6 mol·L^{-1}），NaClO（新配），$FeSO_4$（0.1 mol·L^{-1}），NH_4SCN（饱和），NH_4F（2 mol·L^{-1}），$Fe(NO_3)_3$（0.1 mol·L^{-1}），$CoCl_2$（0.1 mol·L^{-1}），Na_2S（0.1 mol·L^{-1}），KI（0.1 mol·L^{-1}），$NiSO_4$（0.1 mol·L^{-1}），$BaCl_2$（0.5 mol·L^{-1}），$K_3[Fe(CN)_6]$（0.1 mol·L^{-1}），$K_4[Fe(CN)_6]$（0.1 mol·L^{-1}），氯水（新配），溴水，Fe^{3+}、Co^{2+}、Ni^{2+} 混合溶液 4 组，H_2O_2（质量分数为 3%），CCl_4，戊醇，铁粉，丁二酮肟试剂。

【实验内容】

用表格报告实验结果，解释现象时，若涉及化学反应，均要求写出反应的化学方程式。

1. 氢氧化物的生成和性质

（1）$M(OH)_2$ 的生成和性质：观察 $FeSO_4$，$CoCl_2$ 和 $NiSO_4$ 溶液的颜色，检验其 pH，由这 3 种盐溶液和其他合适试剂作用获得 $Fe(OH)_2$、$Co(OH)_2$ 和 $Ni(OH)_2$，观察沉淀的颜色及它们的酸碱性。将沉淀放置一段时间再观察比较其颜色变化。

提示：$Fe(OH)_2$ 极易被溶液中的氧气氧化，必须很小心地操作才能观察到白色 $Fe(OH)_2$ 的生成。制备时，可以在一支试管中加入 5 mL 6 mol·L^{-1} 新配 NaOH 溶液，小心煮沸，冷却备用；用另一支试管取适量 $FeSO_4$ 溶液，用稀硫酸酸化，加入少量铁粉，煮沸，用一细长滴管吸取冷却的 NaOH 溶液，小心插入 $FeSO_4$ 溶液中，缓慢放出 NaOH。

（2）$M(OH)_3$ 的生成和性质：分别由 $FeSO_4$、$CoCl_2$ 和 $NiSO_4$ 溶液和其他合适的试剂作用，获得 $Fe(OH)_3$、$Co(OH)_3$ 和 $Ni(OH)_3$。观察比较它们的状态和颜色，将沉淀离心分离。分别在沉淀中加入浓盐酸，摇荡，微热，并用湿润的 KI$^-$ 淀粉试纸检验是否有氧化性气体逸出。

提示：制备 $M(OH)_3$ 时要注意依据 M(Ⅱ)还原性的强弱选择不同氧化能力的氧化剂，包括空气中的氧、溴水、H_2O_2、氯水、NaOH 溶液等。

2. 硫化物的生成和性质

（1）Fe(Ⅱ)的硫化物：由 $FeSO_4$ 溶液、饱和 H_2S 溶液及其他合适试剂制备 FeS，观察 FeS 的状态和颜色，试验 FeS 与稀盐酸的作用情况。

（2）Co(Ⅱ)、Ni(Ⅱ)的硫化物：分别由 $CoCl_2$、$NiSO_4$ 溶液代替 $FeSO_4$ 做同样的试验。

（3）Fe(Ⅲ)在不同介质中与 S^{2-} 的作用：

① 取适量饱和 H_2S 溶液，逐滴加入 $Fe(NO_3)_3$ 溶液，摇荡，观察沉淀的状态和颜色。

② 用 Na_2S 溶液代替饱和 H_2S 溶液做上述试验。

3. 配合物的生成与性质

（1）氨的配合物：分别向 $Fe(NO_3)_3$，$FeSO_4$，$CoCl_2$ 和 $NiSO_4$ 溶液中滴加 6 mol·L^{-1} 的氨水，观察沉淀的生成，继续滴加氨水观察沉淀是否溶解。比较颜色变化。

（2）与 SCN^- 形成的配合物：用 NH_4SCN 溶液代替氨水做上述试验，观察比较颜色变化。

（3）配合物的生成对氧化还原性质的影响：取少量 $Fe(NO_3)_3$ 溶液，滴加 2~3 滴 CCl_4，逐滴加入 KI 溶液，摇荡，观察 CCl_4 层的颜色变化，向此混合溶液中滴加 NH_4F 溶液，摇荡，观察 CCl_4 层的颜色变化。

（4）配合物的生成在离子鉴定中的应用：

① 与 $K_3[Fe(CN)_6]$ 的作用：在点滴板上，各取少量的 $Fe(NO_3)_3$、$FeSO_4$、$CoCl_2$ 和 $NiSO_4$ 溶液，向各溶液中逐滴加入 $K_3[Fe(CN)_6]$ 试剂。观察并比较状态和颜色变化(本实验可直接在滤纸上做)。

② 与 $K_4[Fe(CN)_6]$ 的作用：用 $K_4[Fe(CN)_6]$ 代替 $K_3[Fe(CN)_6]$ 做上述同样的试验，观察现象，总结 Fe^{3+}、Fe^{2+}、Co^{2+}、Ni^{2+} 4 种离子分别与 $[Fe(CN)_6]^{3-}$ 和 $[Fe(CN)_6]^{4-}$ 作用的情况。

③ 从 Fe^{3+}、Co^{2+}、Ni^{2+} 混合液中检出单一离子：取 1 mL 混合液放入试管中，加入 NH_4SCN 溶液 0.5 mL，观察溶液颜色变化，缓慢加入 NH_4F 溶液，然后加入 1 mL 戊醇，摇荡后静置，观察戊醇层颜色变化；用吸管吸取试管下层(水相)溶液 2～3 滴，放在点滴板上，再滴加 1～2 滴氨水，最后加入 1～2 滴丁二酮肟溶液，观察沉淀的颜色变化。根据试验结果判断 Fe^{3+}、Co^{2+}、Ni^{2+} 的存在。

【思考题】

1. 制备 $Fe(OH)_2$ 时，有关溶液均需煮沸并避免振荡，为什么？
2. 试总结铁、钴、镍硫化物的性质。
3. 怎样从 $Fe(OH)_3$ 制备 $FeCl_2$？写出反应的化学方程式。
4. 比较 $Fe(OH)_3$、$Al(OH)_3$、$Cr(OH)_3$ 的性质。怎样利用这些性质把 Fe^{3+}、Al^{3+}、Cr^{3+} 从溶液中分离出来？

● 视频演示

第 4 章 元素化学实验

风险评估及控制工作表

实验任务		铁、钴、镍的化合物				评估人 吴泽颖,王红松,吴星星				评估日期 2021 年 7 月 1 日				
过程步骤	化合物名称	危害鉴别				健康风险		安全措施			清单/风险评估	风险评估		紧急情况/泼洒措施
		类别	浓度	用量	化学品安全技术说明书	可能的伤害种类	降低或消除风险	技术	防护眼镜	手套	通风橱	可能性 1~4	危害 1~4	风险 1~4
		或物理危害												危害级别 高中低	特殊备注

A: 氢氧化物的生成和性质

I) M(OH)₂ 的生成和性质:观察 FeSO₄、CoCl₂ 和 NiSO₄ 溶液的颜色,检验其pH,由这3种盐溶液和其他合适试剂获得 Fe(OH)₂、Co(OH)₂ 和 Ni(OH)₂,观察沉淀的颜色及它们的酸碱性。将沉淀放置一段时间再观察比较其颜色变化。

	FeSO₄	—	0.1 M	适量	√	皮肤/眼刺激	关注		√	√	√	1	1	1	低	大量水冲洗
	CoCl₂	—	0.1 M	适量	√	皮肤过敏	关注		√	√	√	1	1	1	低	大量水冲洗
	NiSO₄	—	0.1 M	适量	√	皮肤刺激	关注		√	√	√	1	1	1	低	大量水冲洗

II) M(OH)₃ 的生成和性质:分别由 FeSO₄、CoCl₂ 和 NiSO₄ 和其他合适的试剂作用,获得 Fe(OH)₃、Co(OH)₃ 和 Ni(OH)₃。观察比较它们的状态和颜色,分别在沉淀中加入浓盐酸,摇匀,微热。

| | 浓盐酸 | 2.3+8 | 6 M | 适量 | √ | 皮肤灼伤和眼损伤 | 关注 | | √ | √ | √ | 1 | 4 | 4 | 低 | 大量水冲洗 |

B: 硫化物的生成和性质

I) Fe(II) 的硫化物:由 FeSO₄ 溶液、饱和 H₂S 溶液及其他合适试剂制备 FeS,观察 FeS 的状态和颜色,试验 FeS 与稀盐酸的作用情况。

| | 饱和 H₂S 溶液 | 2.3+2.1 | | 适量 | √ | 剧毒 | 关注 | | √ | √ | √ | 1 | 4 | 4 | 低 | 大量水冲洗 |

II) Co(II)、Ni(II) 的硫化物:分别由 CoCl₂、NiSO₄ 溶液代替 FeSO₄ 做同样的试验。

续 表

风险评估及控制工作表

实验任务: 铁、钴、镍的化合物　　评估人: 吴泽颖、王红松、吴星星　　评估日期: 2021年7月1日

过程步骤	化合物名称	危害鉴别			评估风险		安全措施			风险评估			紧急情况/泄洒措施	特殊备注
		类别	浓度	用量	化学品安全技术说明书	健康或可能的伤害种类	降低或消除风险	技术	防护眼镜	手套	通风橱	清单/风险评估	可能性 1~4	危害 1~4	风险	危害级别 高中低
Ⅲ）Fe（Ⅲ）在不同介质中与S²⁻的作用: （1）取适量饱和H₂S溶液，逐滴加入Fe(NO₃)₃溶液，摇荡，观察沉淀的状态和颜色。 （2）用Na₂S溶液代替饱和H₂S溶液做上述试验。	Fe(NO₃)₃	5.1	0.1 M	适量	√	—	关注						1	1	1	低	大量水冲洗
	Na₂S	4.2	0.1 M	适量	√	皮肤灼伤和眼损伤	关注						1	2	2	低	大量水冲洗
C: 配合物的生成与性质																	
Ⅰ）氨的配合物: 分别向Fe(NO₃)₃、FeSO₄、CoCl₂和NiSO₄溶液中滴加6 mol·L⁻¹的氨水，观察沉淀的生成，继续滴加氨水观察沉淀是否溶解。比较颜色。	氨水	8	6 M	适量	√	皮肤灼伤和眼损伤	关注	√	√	√	√		1	3	3	低	大量水冲洗
Ⅱ）与SCN⁻形成的配合物: 用NH₄SCN溶液代替氨水做上述试验。	饱和NH₄SCN溶液	—	—	适量	√	吞咽有害	关注	√	√	√			1	3	3	低	大量水冲洗
Ⅲ）配合物的生成对氧化还原性质的影响: 取少量Fe(NO₃)₃溶液，加入2~3滴CCl₄，逐滴加入KI溶液，摇荡，观察CCl₄层颜色变化，向此混合溶液中滴加NH₄F溶液，摇荡，观察CCl₄层的颜色变化。	CCl₄	6.1		2~3滴	√	吞咽/皮肤接触会中毒	关注	√	√	√			1	4	4	低	大量水冲洗
	KI	—	0.1 M	适量	√	—	关注	√	√	√			1	1	1	低	大量水冲洗
	NH₄F	6.1	2 M	适量	√	吞咽/皮肤接触会中毒	关注	√	√	√			1	2	2	低	大量水冲洗

风险评估及控制工作表

实验任务：铁、钴、镍的化合物　　**评估人**：吴泽颖、王红松、吴星星　　**评估日期**：2021年7月1日

过程步骤	化合物名称	危害鉴别 类别	浓度	用量	化学品安全技术说明书	健康风险 可能的伤害种类	降低或消除风险	安全措施 技术	防护眼镜	手套	通风橱	评估清单/风险评估	风险评估 可能性	危害	风险	危害级别	紧急情况/泼洒措施	特殊备注
Ⅳ）配合物的生成在离子鉴定中的应用： (1) 与$K_3[Fe(CN)_6]$的作用：在点滴板上，各取少量的$Fe(NO_3)_3$、$FeSO_4$、$CoCl_2$和$NiSO_4$溶液，向各溶液中逐滴滴入$K_3[Fe(CN)_6]$试剂。观察溶液颜色变化和颜色变化状态并直接在滤纸上做。（本实验可直接在滤纸上做）	$K_3[Fe(CN)_6]$	—	0.1 M	适量	√	皮肤/眼刺激	关注		√	√			1	2	2	低	大量水冲洗	
(2) 与$K_4[Fe(CN)_6]$的作用：用$K_4[Fe(CN)_6]$代替$K_3[Fe(CN)_6]$做上述同样的试验，观察现象，总结Fe^{3+}、Fe^{2+}、Co^{2+}、Ni^{2+}4种离子分别与$[Fe(CN)_6]^{3-}$和$[Fe(CN)_6]^{4-}$作用的情况。	$K_4[Fe(CN)_6]$	—	0.1 M	适量	√	—	关注		√	√			1	1	1	低	大量水冲洗	
③从Fe^{3+}、Co^{2+}、Ni^{2+}混合液中检出单一离子：取1 mL混合液放入试管中，加入NH_4SCN溶液0.5 mL，观察溶液颜色变化，缓慢加入NH_4F溶液，然后加入1 mL戊醇，摇荡后静置，观察戊醇层颜色变化；用吸管吸取下层（水相）溶液2~3滴，放在点滴板上，再滴加1~2滴氨水，观察沉淀的颜色变化。根据试验结果判断Fe^{3+}、Co^{2+}、Ni^{2+}的存在。	饱和NH_4SCN溶液	—	—	适量	√	吞咽有害	关注	√	√				1	3	3	低	大量水冲洗	
	丁二酮肟	—	—	1~2滴	√	吞咽会中毒	关注		√	√			1	2	2	低	大量水冲洗	
	戊醇	3.3	—	1 mL	√	皮肤/眼刺激	关注		√	√	√		1	3	3	低	大量水冲洗	

续 表

风险评估及控制工作表

实验任务	铁、钴、镍的化合物				评估人	吴泽颖、王红松、吴星星				评估日期	2021 年 7 月 1 日		
过程步骤	危害鉴别				健康风险	安全措施				清单/风险评估	风险评估		措施
	类别	浓度	用量	化学品安全技术说明书	可能的伤害种类	降低或消除风险	技术	防护眼镜	手套		可能性	危害 风险	危害级别	紧急情况/泼洒	特殊备注

化合物名称	类别	浓度	用量	化学品安全技术说明书	可能的伤害种类	降低或消除风险	技术	防护眼镜	手套	通风橱	可能性	危害	风险	危害级别	紧急情况/泼洒	特殊备注
				或物理危害							1~4	1~4		高中低		
Fe(OH)$_3$	—			√	—	关注		√	√		1	1	1	低	—	—
Co(OH)$_3$	—			√	—	关注		√	√		1	1	1	低	—	—
Ni(OH)$_3$	—			√	—	关注		√	√		1	1	1	低	—	—

实验产物

实验废弃物

整个过程的最终评估　低风险（按照正确流程操作）

签名　实验老师：　　　　实验室主任/职业安全健康（OHS）代表：　　　　学生/操作人：

Experiment 10 Compounds of Fe, Co and Ni

【Objectives】

1. To master the formation and properties of iron, cobalt and nickel hydroxides.
2. To master the formation and properties of iron, cobalt and nickel sulphides.
3. To master the formation and properties of iron, cobalt and nickel complexes.

【Principles】

Fe, Co and Ni have similar configurations ($[Ar]3d^{6\sim 8}4s^2$) and similar atomic radii (115~117 pm).

The salts of Fe(Ⅲ), Fe(Ⅱ), Co(Ⅱ) and Ni(Ⅱ) are common. Their nitrates, sulfates and halides are all soluble in water. The hydration of these cations results in color changes. For example, Fe^{2+} changes from white to light green; Co^{2+} changes from blue to pink; Ni^{2+} changes from yellow to bright green. These salts usually exist as a series of compounds that differ in their degree of hydration, and have different colors. For example:

$$CoCl_2 \cdot 6H_2O \xrightleftharpoons{520\ ℃} CoCl_2 \cdot 2H_2O \xrightleftharpoons{90\ ℃} CoCl_2 \cdot H_2O \xrightleftharpoons{120\ ℃} CoCl_2$$
$$\text{(pink)} \qquad\qquad \text{(red purple)} \qquad\qquad \text{(purple blue)} \qquad\qquad \text{(blue)}$$

This property of $CoCl_2$ is usually used in the laboratory to show the effects of desiccants (silica gel).

The salts of Fe^{3+} are similar to that of Cr^{3+} and Al^{3+}. For example, MCl_3 can form dimer M_2Cl_6, with tetracoordinate M. Their sulfates and the sulfates of alkali metals can form alums with the general formula $M_2SO_4 \cdot M_2(SO_4)_3 \cdot 24H_2O$. For example, $K_2SO_4 \cdot Al_2(SO_4)_3 \cdot 24H_2O$ and $K_2SO_4 \cdot Fe_2(SO_4)_3 \cdot 24H_2O$.

Table 10-1 shows some important insoluble salts of Fe, Co and Ni.

Table 10-1 Common insoluble salts of Fe, Co and Ni

Cations	Anions			
	S^{2-}	CO_3^{2-}	CrO_4^{2-}	$[Fe(CN)_6]^{4-}$
Fe^{3+}	black	—	—	blue
Fe^{2+}	black	white	white	white
Co^{2+}	black	pink	light pink	green
Ni^{2+}	black	light green	light green	light green

The cations of Fe, Co and Ni can form a lot of coordination compounds. Fe^{3+} and Fe^{2+} react strongly with OH^-. So they can't form stable complex ions in ammonia water. Table 10-2 shows some common and stable coordination compounds of Fe, Co and Ni. These coordination compounds have different solubilities and colors which can be used to identify Fe, Co and Ni.

Table 10-2 Common coordination compounds of Fe, Co and Ni

Metal ions	Ligands				
	H_2O	CN^-	NH_3	SCN^-	F^-
Fe^{3+}	light purple	light yellow	—	blood red	colorless
Fe^{2+}	light green	yellow	—	colorless	—
Co^{3+}	—	—	red	—	—
Co^{2+}	pink	purple	orange	blue	—
Ni^{2+}	green	yellow	Dark blue	—	—

The acidic and basic strength of the hydroxides of Fe, Co and Ni abides by the general rules as follows:

Because the hydroxides of M(Ⅵ) are strong acids, their general formula is MO_4^{2-} no matter in acidic or basic solutions. The color of FeO_4^{2-} is purple, the same as MnO_4^-. Similar to MnO_4^-, CrO_4^{2-} and SO_4^{2-}, FeO_4^{2-} can precipitate Ba^{2+}.

The hydroxide formula for Fe, Co and Ni is $M(OH)_3$ or $M(OH)_2$, and they are not strong bases. Freshly prepared $Fe(OH)_3$ is weakly acidic, and can be dissolved in hot concentrated KOH solution to form $KFeO_2$. The soluble salts of Fe, Co and Ni all hydrolyze. The hydrolysis of M^{3+} is stronger than that of M^{2+}, especially for Fe^{3+}. The hydrolysis products of Fe^{3+} will make the solution brown or red brown.

$$Fe^{3+} + nH_2O \rightleftharpoons [Fe(OH)_n]^{(3-n)} + nH^+ \quad (n=1, 2, 3)$$

Fe^{3+} can only exist in strong acid solutions (pH<2). Diluting the solution or increasing the pH value of the solution can produce red brown $Fe(OH)_3$ precipitate.

The standard potentials $E_A^{\ominus}/V$ for compounds of Fe, Co and Ni are as follows (data or compounds in square brackets are the standard potentials $E_B^{\ominus}/V$ or compounds in the basic solution):

$$FeO_4^{2-} \xrightarrow[[+0.9]]{+1.9} \underset{[Fe(OH)_3]}{Fe^{3+}} \xrightarrow[[-0.56]]{+0.771} \underset{[Fe(OH)_2]}{Fe^{2+}} \xrightarrow[[-0.887]]{-0.440} Fe$$

$$CoO_2 \xrightarrow[[+0.7]]{\geq +1.8} \underset{[Co(OH)_3]}{Co^{3+}} \xrightarrow[[+0.17]]{+1.82} \underset{[Co(OH)_2]}{Co^{2+}} \xrightarrow[[-0.73]]{-0.277} Co$$

$$NiO_4^{2-} \xrightarrow[[>+0.4]]{+1.8} NiO_2 \xrightarrow[[+0.49]]{1.68} \underset{[Ni(OH)_2]}{Ni^{2+}} \xrightarrow[[-0.72]]{-0.250} Ni$$

For the compounds of Fe, Co and Ni, if the oxidation numbers of Fe, Co and Ni are equal to or greater than +4, they are very strong oxidants. In acidic solutions, their oxidizing strength is greater than MnO_4^-. But in basic solutions, their oxidizing strength will be greatly reduced. The compounds of Fe, Co and Ni, which have small oxidation numbers, can be oxidized by strong oxidants, such as NaClO, in strong bases. For example:

$$2Fe(OH)_3 + 3ClO^- + 4OH^- = 2FeO_4^{2-} + 3Cl^- + 5H_2O$$

$$Fe_2O_3 + 3KNO_3 + 4KOH \xrightarrow{melting} 2K_2FeO_4 + 3KNO_2 + 2H_2O$$

M(Ⅲ) is an oxidant in acidic solutions. Fe^{3+} is a medium strong oxidant. Co^{3+} is a strong oxidant, and can oxidize H_2O_2 and HCl. Ni^{3+} is the strongest oxidant of the three. In acidic solutions, Fe^{3+} can be reduced by Fe, Cu, Sn^{2+}, S^{2-}, I^- and SO_2.

M(Ⅲ) can form stable coordination compounds and insoluble salts. For example:

$$Fe^{3+} \xrightarrow{+0.771} Fe^{2+} \qquad\qquad Co^{3+} \xrightarrow{+1.842} Co^{2+}$$

$$[FeF_6]^{3-} \xrightarrow{-0.135} Fe^{2+} \qquad\qquad [Co(NH_3)_6]^{3+} \xrightarrow{+0.1} [Co(NH_3)_6]^{2+}$$

$$[Fe(CN)_6]^{3-} \xrightarrow{+0.36} [Fe(CN)_6]^{4-} \qquad [Co(CN)_6]^{3-} \xrightarrow{+0.81} [Co(CN)_6]^{4-}$$

For Fe, Co and Ni, the redox properties of M(Ⅲ), M(Ⅱ) and M(0), and the acidic or basic strength of their hydroxides are generalized in the following diagrams.

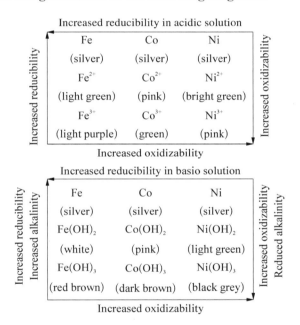

【Identification of Fe, Co and Ni】

(1) Fe^{3+} In weak acid solution, Fe^{3+} reacts with KSCN to form blood red complex ion $[Fe(SCN)_n]^{(3-n)}$.

$$Fe^{3+} + nKSCN \Longrightarrow [Fe(SCN)_n]^{(3-n)} + nK^+$$

Other ions won't interfere with this reaction. If the concentrations of heavy metal ions are not high, Fe^{3+} reacts with $K_4[Fe(CN)_6]$ to from a dark blue complex called prussian blue $[KFe(CN)_6Fe]_x$:

$$xFe^{3+} + xK^+ + x[Fe(CN)_6]^{4-} \Longrightarrow [KFe(CN)_6Fe]_x \downarrow$$

(2) Fe^{2+} Fe^{2+} reacts with $K_3[Fe(CN)_6]$ to form dark blue $[KFe(CN)_6Fe]_x$.

$$xFe^{2+} + xK^+ + x[Fe(CN)_6]^{3-} \Longrightarrow [KFe(CN)_6Fe]_x \downarrow$$

If the concentrations of heavy metal ions are not high, this reaction is both high sensitivity and selectivity.

(3) Co^{2+} Co^{2+} reacts with KSCN to form blue $[Co(SCN)_4]^{2-}$

$$Co^{2+} + 4SCN^- = [Co(SCN)_4]^{2-}$$

This complex ion is stable in acetone, so acetone or pentanol should be added when identifying Co^{2+}. If Fe is present, it can be masked by reacting with NaF to form colorless $[FeF_6]^{3-}$.

(4) Ni^{2+} With the presence of $NH_3 \cdot H_2O$, Ni^{2+} reacts with dimethylglyoxime to form a bright red chelate. This is a sensitive reaction to identify Ni^{2+}.

$$Ni^{2+} + 2 \begin{array}{c} CH_3-C=NOH \\ | \\ CH_3-C=NOH \end{array} + 2NH_3 = \begin{array}{c} \text{[Ni dimethylglyoxime chelate]} \end{array} + 2NH_4^+$$

【Apparatus and Chemicals】

1. Apparatus

Starch test-KI paper.

2. Chemicals

HCl (concentrated); H_2S (saturated, freshly prepared); NaOH (s, 6 mol·L^{-1}, freshly prepared); H_2SO_4 (dilute); ammonia water (6 mol·L^{-1}); NaClO (freshly prepared); $FeSO_4$ (0.1 mol·L^{-1}); NH_4SCN (saturated); NH_4F (2 mol·L^{-1}); $Fe(NO_3)_3$ (0.1 mol·L^{-1}); $CoCl_2$ (0.1 mol·L^{-1}); Na_2S (0.1 mol·L^{-1}); KI (0.1 mol·L^{-1}); $NiSO_4$ (0.1 mol·L^{-1}); $BaCl_2$ (0.5 mol·L^{-1}); $K_3[Fe(CN)_6]$ (0.1 mol·L^{-1}); $K_4[Fe(CN)_6]$ (0.1 mol·L^{-1}); chlorine water (freshly prepared); bromine water; four sets of Fe^{3+}、Co^{2+}、Ni^{2+} mixed solutions; H_2O_2 (3%, wt/wt); CCl_4; pentanol; iron powder; dimethylglyoxime reagent.

【Experimental Procedures】

Fill in the appropriate table and write the chemical equations in your report.

1. Formation and properties of hydroxide

(1) Formation and properties of $M(OH)_2$: Observe the colors of $FeSO_4$ solution, $COCl_2$ solution and $NiSO_4$ solution, and test their pH values. Use the above solutions and other appropriate reagents to prepare $Fe(OH)_2$, $Co(OH)_2$ and $Ni(OH)_2$. Observe the colors of the precipitates and test their pH values. After a while, observe the precipitates again and compare.

Tips: $Fe(OH)_2$ is easily oxidized by oxygen. So the formation of white $Fe(OH)_2$ must be performed very carefully. Place 5 mL of freshly prepared 6 mol L^{-1} NaOH into a test tube, then heat and boil carefully. Let the solution cool and it will be ready to use. Put an appropriate amount of $FeSO_4$ solution in another test tube, and add dilute H_2SO_4 and a small amount of iron powder. Then, boil the solution. Use a long thin pipet to draw NaOH, and put the tip of the pipet into $FeSO_4$ solution carefully. Then release NaOH slowly.

(2) Formation and properties of M(OH)$_3$: Use FeSO$_4$ solution, CoCl$_2$ solution and NiSO$_4$ solution, and other appropriate reagents to prepare Fe(OH)$_3$, Co(OH)$_3$ and Ni(OH)$_3$. Observe and compare their states and colors. Centrifuge to get the precipitates. Add concentrated HCl to each precipitate, and heat gently. Test if gases are produced with moistened KI-starch test paper.

Tips: According to the reductivity of M(II), choose appropriate oxidants to prepare M(OH)$_3$, such as oxygen in the air, bromine water, H$_2$O$_2$, chlorine water, NaOH solution, etc.

2. Formation and properties of sulfides

(1) Sulfides of Fe(II): Use FeSO$_4$ solution, saturated H$_2$S solution and other appropriate reagents to prepare FeS. Observe the state and the color of FeS. Evaluate the reaction of FeS with dilute HCl

(2) Sulfides of Co(II) and Ni(II): Replace FeSO$_4$ solution with CoCl$_2$ solution and NiSO$_4$ solution to replicate the above experiments.

(3) Reactions of Fe(III) and S^{2-} in different media:

① Add Fe(NO$_3$)$_3$ solution drop by drop to an appropriate amount of saturated H$_2$S solution. Shake well and observe the precipitate.

② Use Na$_2$S solution to replace saturated H$_2$S solution to replicate the above experiment.

3. Formation and properties of coordination compounds

(1) Coordination compounds containing ammonia: Add 6 mol·L^{-1} ammonia water to Fe(NO$_3$)$_3$, FeSO$_4$, CoCl$_2$ and NiSO$_4$ respectively. Observe the formation of precipitates. Continue adding ammonia water, and observe whether the precipitates dissolve. Compare the color changes.

(2) Coordination compounds formed with SCN$^-$: Replace ammonia water with NH$_4$SCN solution to replicate the above experiments, and compare the color changes.

(3) Effects of the formation of coordination compounds on redox properties: Add 2 to 3 drops of CCl$_4$ to a small amount of Fe(NO$_3$)$_3$ solution, then add KI solution drop by drop. Shake well and observe the color of the CCl$_4$ layer. Continue adding NH$_4$F solution. Shake well and observe the color of the CCl$_4$ layer again.

(4) Identification of ions by coordination reactions:

① Reaction with K$_3$[Fe(CN)$_6$]: Add Fe(NO$_3$)$_3$, FeSO$_4$, CoCl$_2$ and NiSO$_4$ solutions on a spot plate respectively. Then add K$_3$[Fe(CN)$_6$] drop by drop to the four solutions. Observe and compare the states and colors. (The experiments can be performed on filter paper.)

② Reaction with K$_4$[Fe(CN)$_6$]: Use K$_4$[Fe(CN)$_6$] to replace K$_3$[Fe(CN)$_6$] to replicate the above experiments. Summarize the reactions of four ions, Fe^{3+}, Fe^{2+}, Co^{2+} and Ni^{2+}, with [Fe(CN)$_6$]$^{3-}$ and [Fe(CN)$_6$]$^{4-}$ respectively.

③ Identification of Fe^{3+}, Co^{2+} and Ni^{2+} in a mixed solution: Transfer 1 mL of mixed solution in a test tube, and add 0.5 mL of NH$_4$SCN solution. Observe the color of the solution. Then slowly add NH$_4$F solution and 1 mL of pentanol. Shake well and observe the color of the pentanol layer. Use a pipet to draw 2 to 3 drops of the solution from the lower layer (aqueous phase) onto a spot plate. Next add 1 to 2 drops of ammonia water, and 1 to 2 drops of dimethylglyoxime solution. Observe the color change of the precipitate. Identify the presence of Fe^{3+}, Co^{2+} and Ni^{2+} according to the results of the experiments.

【Questions】

1. When preparing $Fe(OH)_2$, the solutions should be boiled and shaking should be avoided. Why?

2. To identify Co^{2+} in this experiment, why should NH_4F solution and pentanol be added? When is NH_4F not needed? What is the blue compound in the pentanol layer? Does the formation of the blue compound occur before or after the pentanol is added?

3. How do we prepare $Fe(OH)_3$ from $FeCl_2$? Record the chemical equations.

4. Compare the properties of $Fe(OH)_3$, $Al(OH)_3$ and $Cr(OH)_3$. How do we separate Fe^{3+}, Al^{3+} and Cr^{3+} by their properties?

Risk Assessment and Control Worksheet

TASK	Compounds of Fe, Co and Ni				Assessor	Zeying WU, Hongsong WANG, Xingxing WU				Date	1 JUL. 2021			
Process Steps	Hazard Identification				Health Risk	Safety Measures				Checklist＼Risk Assessment	Risk Assessment			Action
	Chemical Name	Class	Concn.	Amount	Type of Injury Possible	To reduce or eliminate the risk	Technique	Glasses	Gloves	Fume Cbd.	Likely	Hazard	Risk Score	Risk Level	In case of Emergency/Spill
		or Physical Hazard												H M L	Special Comments
A: Formation and properties of hydroxide													1~4	1~4	
I) Formation and properties of M(OH)₂: Observe the colors of FeSO₄ solution, CoCl₂ solution and NiSO₄ solution, and test their pH values. Use the above solutions and other appropriate reagents to prepare Fe(OH)₂, Co(OH)₂ and Ni(OH)₂.	FeSO₄	—	0.1 M	appropriate amount	√	Skin and eye irritant	Handle with care		√	√		1	1	1	L	Sponge & water
	CoCl₂	—	0.1 M	appropriate amount	√	Skin allergy	Handle with care		√	√		1	1	1	L	Sponge & water
	NiSO₄	—	0.1 M	appropriate amount	√	Skin irritant	Handle with care		√	√		1	1	1	L	Sponge & water
II) Formation and properties of M(OH)₃: Centrifuge to get the precipitates. Add concentrated HCl to each precipitate, and heat gently.	HCl	2.3+8	6 M	appropriate amount	√	Skin burns and eye injuries	Handle with care		√	√	√	1	4	4	L	Sponge & water
B: Formation and properties of sulfides																
I) Sulfides of Fe(II): Use FeSO₄ solution, saturated H₂S solution and other appropriate reagents to prepare FeS. Observe the state and the color of FeS. Evaluate the reaction of FeS with dilute HCl.	Saturated H₂S solution	2.3+2.1		appropriate amount	√	Toxic	Handle with care		√	√	√	1	4	4	L	Sponge & water
II) Sulfides of Co(II) and Ni(II): Replace FeSO₄ solution with CoCl₂ solution and NiSO₄ solution to replicate the above experiments.																

continued

Risk Assessment and Control Worksheet

TASK	Compounds of Fe, Co and Ni						Assessor	Zeying WU, Hongsong WANG, Xingxing WU				Date				1 JUL. 2021	
Process Steps	Hazard Identification					Health Risk		Safety Measures				Checklist～Risk Assessment				Action	
	Chemical Name	Class	Concn.	Amount	MSDS	Type of Injury Possible	To reduce or eliminate the risk	Technique	Glasses	Gloves	Fume Cbd.	Likely	Hazard	Risk Score	Risk Level	In case of Emergency/Spill	Special Comments
		or Physical Hazard										1～4	1～4		H M L		
Ⅲ) Reactions of Fe(Ⅲ) and S^{2-} in different media: (1) Add $Fe(NO_3)_3$ solution drop by drop to an appropriate amount of saturated H_2S solution. Shake well and observe the precipitate.	$Fe(NO_3)_3$	5.1	0.1 M	appropriate amount	✓	—	Handle with care		✓	✓		1	1	1	L	Sponge & water	
(2) Use Na_2S solution to replace saturated H_2S solution to replicate the above experiment.	Na_2S	4.2	0.1 M	appropriate amount	✓	Skin burns and eye injuries	Handle with care		✓	✓		1	2	2	L	Sponge & water	
C: Formation and properties of coordination compounds																	
Ⅰ) Coordination compounds containing ammonia: Add $6\ mol \cdot L^{-1}$ ammonia water to $Fe(NO_3)_3$, $FeSO_4$, $CoCl_2$ and $NiSO_4$ respectively. Observe the formation of precipitates. Continue adding ammonia water, and observe whether the precipitates dissolve. Compare the color changes.	Ammonia	8	6 M	Appropriate amount	✓	Skin burns and eye injuries	Handle with care		✓	✓	✓	1	3	3	L	Sponge & water	
Ⅱ) Coordination compounds formed with SCN^-: Replace ammonia water with NH_4SCN solution to replicate the above experiments, and compare the color changes.	Saturated NH_4SCN solution	—		appropriate amount	✓	Harmful if swallowed	Handle with care		✓	✓	✓	1	3	3	L	Report immediately to Demonstrator or Prep Room	

Risk Assessment and Control Worksheet (continued)

TASK: Compounds of Fe, Co and Ni   Assessor: Zeying WU, Hongsong WANG, Xingxing WU   Date: 1 JUL. 2021

Process Steps	Chemical Name	Hazard Identification — Class	Concn.	Amount	MSDS	Health Risk — Type of Injury Possible	To reduce or eliminate the risk	Safety Measures — Technique	Glasses	Gloves	Fume Cbd.	Checklist—Risk Assessment	Risk Assessment — Likely	Hazard	Risk Score	Risk Level H M L	In case of Emergency/Spill	Special Comments
Ⅲ) Effects of the formation of coordination compounds on redox properties: Add 2 to 3 drops of CCl_4 to a small amount of $Fe(NO_3)_3$ solution, then add KI solution drop by drop. Shake well and observe the color of the CCl_4 layer. Continue adding NH_4F solution. Shake well and observe the color of the CCl_4 layer again.	CCl_4	6.1		2~3 drops	√	Swallowing/ skin contact can be toxic	Handle with care		√	√	√		1	1~4	4	L	Sponge & water	
	KI	—	0.1 M	appropriate amount	√	—	Handle with care		√	√			1	1	1	L	Sponge & water	
	NH_4F	6.1	2 M	appropriate amount	√	Swallowing/ skin contact can be toxic	Handle with care		√	√			1	2	2	L	Sponge & water	
Ⅳ) Identification of ions by coordination reactions: (1) Reaction with $K_3[Fe(CN)_6]$: Add $Fe(NO_3)_3$, $FeSO_4$, $CoCl_2$ and $NiSO_4$ solutions on a spot plate respectively. Then add $K_3[Fe(CN)_6]$ drop by drop to the four solutions. Observe and compare the states and colors. (The experiments can be performed on filter paper.)	$K_3[Fe(CN)_6]$		0.1 M	appropriate amount	√	Skin and eye irritant	Handle with care		√	√			1	2	2	L	Sponge & water	
(2) Reaction with $K_4[Fe(CN)_6]$: Use $K_4[Fe(CN)_6]$ to replace $K_3[Fe(CN)_6]$ to replicate the above experiments. Summarize the reactions of four ions, Fe^{3+}, Fe^{2+}, Co^{2+} and Ni^{2+}, with $[Fe(CN)_6]^{3-}$ and $[Fe(CN)_6]^{4-}$ respectively.	$K_4[Fe(CN)_6]$	—	0.1 M	appropriate amount	√	—	Handle with care		√	√			1	1	1	L	Sponge & water	

continued

Risk Assessment and Control Worksheet

TASK	Compounds of Fe, Co and Ni					Assessor	Zeying WU, Hongsong WANG, Xingxing WU					Date	1 JUL. 2021		
Process Steps	Hazard Identification					Health Risk	Safety Measures				Checklist~Risk Assessment			Action	
	Chemical Name	Class	Concn.	Amount	MSDS	Type of Injury Possible	To reduce or eliminate the risk	Technique	Glasses	Gloves	Fume Cbd.	Likely 1~4	Hazard 1~4	Risk Score	Risk Level H M L	In case of Emergency/Spill	Special Comments
(3) Identification of Fe^{3+}, Co^{2+} and Ni^{2+} in a mixed solution: Transfer 1 mL of mixed solution in a test tube, and add 0.5 mL of NH_4SCN solution. Observe the color of the solution. Then slowly add NH_4F solution and 1 mL of pentanol. Shake well and observe the color of the pentanol layer. Use a pipet to draw 2 to 3 drops of the solution from the lower layer (aqueous phase) onto a spot plate. Next add 1 to 2 drops of ammonia water, and 1 to 2 drops of dimethylglyoxime solution. Observe the color change of the precipitate. Identify the presence of Fe^{3+}, Co^{2+} and Ni^{2+} according to the results of the experiments.	Saturaed NH_4SCN Solution	—	or Physical Hazard	appropriate amount	√	Harmful if swallowed	Handle with care		√	√		1	3	3	L	Report immediately to Demonstrator or Prep Room
	Dimethylglyoxime	—		1~2 drops	√	Swallowing can be toxic	Handle with care		√	√		1	2	2	L	Report immediately to Demonstrator or Prep Room
	Pentanol	3.3		1 mol	√	Skin and eye irritant	Handle with care		√	√	√	1	3	3	L	Sponge & water
Experimental Product (s)	$Fe(OH)_3$	—			√	—	Handle with care		√	√		1	1	1	L	—
	$Co(OH)_3$	—			√	—	Handle with care		√	√		1	1	1	L	—
	$Ni(OH)_3$	—			√	—	Handle with care		√	√		1	1	1	L	—
Experimental Waste (s)																
Final assessment of overall process	Low risk when correct procedures followed															
Signatures	Supervisor:					Lab Manager/OHS Representative:					Student/Operator:				

第 5 章
物质的提纯与分析

实验 11　氯化钠的提纯

【实验目的】

1. 掌握提纯 NaCl 的原理和方法。
2. 学习过滤、蒸发、浓缩、结晶等基本操作。
3. 了解 SO_4^{2-}、Ca^{2+}、Mg^{2+} 等离子的定性鉴定。

【实验原理】

化学试剂或医药用的 NaCl 都是以粗食盐为原料提纯的,粗食盐中含有 SO_4^{2-}、K^+、Ca^{2+}、Mg^{2+} 等可溶杂质和泥沙等不溶杂质。选择适当的试剂可使 SO_4^{2-}、Ca^{2+}、Mg^{2+} 等离子生成沉淀而除去。一般做法是先在食盐溶液中加入 $BaCl_2$ 溶液,除去 SO_4^{2-}。

$$Ba^{2+} + SO_4^{2-} == BaSO_4 \downarrow$$

然后在溶液中加入 Na_2CO_3 溶液,除去 Ca^{2+}、Mg^{2+} 和过量的 Ba^{2+}。

$$Ca^{2+} + CO_3^{2-} == CaCO_3 \downarrow$$

$$4Mg^{2+} + 5CO_3^{2-} + 2H_2O == Mg(OH)_2 \cdot 3MgCO_3 \downarrow + 2HCO_3^-$$

$$Ba^{2+} + CO_3^{2-} == BaCO_3 \downarrow$$

过量的 Na_2CO_3 溶液用盐酸中和。粗食盐中的 K^+ 与这些沉淀剂不起作用,仍留在溶液中,由于 KCl 的溶解度比 NaCl 大,而且在粗食盐中的含量较少,所以在蒸发浓缩食盐溶液时 NaCl 结晶出来,而 KCl 仍留在溶液中。

【器材与试剂】

1. 器材:托盘天平,烧杯,酒精灯,石棉网,三角漏斗,铁架台,抽滤瓶,布氏漏斗,真空水泵,蒸发皿,量筒,坩埚钳,pH 试纸。

2. 试剂:HCl(6 mol·L^{-1});HAc(2 mol·L^{-1});NaOH(6 mol·L^{-1});BaCl$_2$(0.5 mol·L^{-1});Na$_2$CO$_3$(饱和);(NH$_4$)$_2$C$_2$O$_4$(饱和);对硝基偶氮间苯二酚(镁试剂)和粗食盐等。

【实验内容】

1. 粗食盐的溶解

称取 3 g 粗食盐于 100 mL 烧杯中,加 20 mL 水,加热搅拌使粗食盐溶解(不溶性杂质沉于底部)。

2. 除去 SO$_4^{2-}$

将溶液加热至沸腾,边搅拌边滴加 BaCl$_2$ 溶液至 SO$_4^{2-}$ 沉淀完全。继续加热煮沸数分钟,抽滤。

3. 除去 Ca^{2+}、Mg^{2+}、Ba^{2+} 等阳离子

将所得的滤液加热到略沸腾,滴加饱和 Na$_2$CO$_3$ 溶液并不断搅拌,直至不再产生沉淀为止。再多加 0.5 mL Na$_2$CO$_3$ 溶液,继续加热 5 min 后,抽滤,弃去沉淀。

4. 除去过量 CO$_3^{2-}$

往溶液中滴加 6 mol·L^{-1} HCl,加热搅拌,调节到溶液的 pH 约为 2~3。

5. 浓缩与结晶

把溶液倒入预先称好的蒸发皿中,蒸发浓缩到有大量 NaCl 结晶出现(约为原来体积的 1/4)。冷却,抽滤。将氯化钠晶体转移到蒸发皿中,小火烘干。冷却后称量,计算产率。

6. 产品纯度的检验

取产品和原料各 1 g,分别溶于 5 mL 蒸馏水中,然后进行下列离子的定性检验。

(1) SO$_4^{2-}$ 各取溶液 1 mL 于试管中,分别加入 6 mol·L^{-1} HCl 溶液 2 滴和 0.5 mol·L^{-1} BaCl$_2$ 溶液 2 滴。比较两溶液中沉淀产生的情况。

(2) Ca^{2+} 各取溶液 1 mL,加 2 mol·L^{-1} HAc 使呈酸性,再分别加入饱和(NH$_4$)$_2$C$_2$O$_4$ 溶液 3~4 滴,若有白色 CaC$_2$O$_4$ 沉淀产生,表示有 Ca^{2+} 存在(该反应可作为 Ca^{2+} 的定性鉴定)。比较两溶液中沉淀产生的情况。

(3) Mg^{2+} 各取溶液 1 mL,加 6 mol·L^{-1} NaOH 溶液 5 滴和镁试剂 2 滴,若有天蓝色沉淀生成,表示有 Mg^{2+} 存在(该反应可作为 Mg^{2+} 的定性鉴定)。比较两溶液的颜色。

【思考题】

1. 为什么用 BaCl$_2$(毒性很大)而不用 CaCl$_2$ 除去 SO$_4^{2-}$?

2. 加 HCl 除 CO$_3^{2-}$ 时,为什么要把溶液的 pH 调到 2~3?调至恰为中性好不好?(提示:从溶液中 H$_2$CO$_3$、HCO$_3^-$ 和 CO$_3^{2-}$ 浓度的比值与 pH 的关系去考虑)

3. 在除去杂质阳离子时,能否用其他可溶性碳酸盐代替 Na$_2$CO$_3$?

· 视频演示

风险评估及控制工作表

实验任务：氯化钠的提纯

评估人： 吴泽颖、王红松、吴星星

评估日期： 2021 年 7 月 1 日

实验任务 过程步骤	危害鉴别 化合物名称	类别	浓度	用量	化学品安全技术说明书或物理危害	健康风险 可能的伤害种类	降低或消除风险	安全措施 技术	防护眼镜	手套	通风橱	评估清单/风险评估	风险评估 可能性 1~4	危害 1~4	风险	危害级别 高中低	紧急情况/泼洒措施	特殊备注
A:粗食盐的溶解 称取 3 g 粗食盐于 100 mL 烧杯中，加 20 mL 水，加热搅拌使粗食盐溶解(不溶性杂质沉于底部)。	酒精灯					烧伤	关注						1	3	3	低	向实验老师或设备准备实验员报告	
B:除去 SO_4^{2-} 将溶液加热至沸腾，边搅拌边滴加 $BaCl_2$ 溶液至 SO_4^{2-} 沉淀完全，继续加热煮沸数分钟，抽滤。	氯化钡	6.1	0.5 M	5 mL	√	吞咽会中毒	关注	√	√	√			1	2	2	低	向实验老师或设备准备实验员报告	
C:除去 Ca^{2+}、Mg^{2+}、Ba^{2+} 等阳离子 将所得的滤液加热到略沸腾，滴加饱和 Na_2CO_3 溶液并不断搅拌，直至不再产生沉淀为止。再多加 0.5 mL Na_2CO_3 溶液，继续加热 5 min 后，抽滤，弃去沉淀。	饱和 Na_2CO_3 溶液	—		少量	√	眼刺激	关注		√	√			1	1	1	低	大量水冲洗	
D:除去过量 CO_3^{2-} 往溶液中滴加 6 mol·L⁻¹ HCl，加热搅拌，中和到溶液的 pH 约为 2~3。	盐酸	2.3+8	6 M	1 mL	√	皮肤灼伤和眼损伤	关注		√	√			1	3	3	低	大量水冲洗	
E:浓缩与结晶																		
F:产品纯度的检验																		

续表

风险评估及控制工作表

评估人：吴泽颖，王红松，吴星星
评估日期：2021年7月1日

实验任务：氯化钠的提纯

过程步骤	化合物名称	危害鉴别 类别	浓度	用量	化学品安全技术说明书	健康风险 可能的伤害种类	降低或消除风险	安全措施 技术	防护眼镜	手套	通风橱	清单/风险评估	风险评估 可能性 1~4	危害 1~4	风险	危害级别 高中低	措施 紧急情况/泼洒	特殊备注
I) 各取溶液 1 mL 于试管中，分别加入 6 mol·L⁻¹ HCl 溶液 2 滴和 0.5 mol·L⁻¹ BaCl₂ 溶液 2 滴。比较两溶液中沉淀产生的情况。																		
II) 各取溶液 1 mL，加 2 mol·L⁻¹ HAc 使呈酸性，再分别加入饱和 (NH₄)₂C₂O₄ 溶液 3~4 滴，若有白色 CaC₂O₄ 沉淀产生，表示有 Ca²⁺ 存在（该反应可作为 Ca²⁺ 的定性鉴定）。比较两溶液中沉淀产生的情况。	HAc	8+3	2 M	1 mL	√	皮肤灼伤和眼损伤	关注		√	√	√		1	1	1	低	用大量水冲洗	
	饱和 (NH₄)₂C₂O₄ 溶液	—		3~4 滴	√	眼刺激	关注		√	√	√		1	3	3	低	大量水冲洗	
III) 各取溶液 1 mL，加 6 mol·L⁻¹ NaOH 溶液 5 滴和镁试剂 2 滴，若有天蓝色沉淀生成，表示有 Mg²⁺ 存在。	氢氧化钠	8	6 M	0.1 mL	√	皮肤灼伤和眼损伤	关注		√	√	√		1	3	3	低	大量水冲洗	
	对硝基偶氮间苯二酚（镁试剂）	—		2 滴	√	皮肤/眼刺激	关注		√	√	√		1	1	1	低	大量水冲洗	

实验产物：

实验废弃物：

整个过程的最终评估：低风险（按照正确流程操作）

签名： 实验老师： 实验室主任/职业安全健康(OHS)代表： 学生/操作人：

Experiment 11 Purification of sodium chloride

【Objectives】

1. To learn the principle and method of purifying sodium chloride.
2. To learn the operation of filtration, evaporation, concentration and crystallization.
3. To learn the methods of qualitative test for SO_4^{2-}, Ca^{2+} and Mg^{2+}.

【Introduction】

Sodium chloride, which is used as a chemical or medical reagent, is purified from crude salt. There are not only insoluble impurities in the crude salt, such as sediment, but also soluble impurities, such as SO_4^{2-}, K^+, Ca^{2+} and Mg^{2+}. To remove Ca^{2+}, Mg^{2+} and SO_4^{2-}, add appropriate reagents to produce insoluble precipitates.

First, add $BaCl_2$ to the crude salt solution to remove SO_4^{2-}.

$$Ba^{2+} + SO_4^{2-} =\!=\!= BaSO_4 \downarrow$$

Then add Na_2CO_3, to remove Ca^{2+}、Mg^{2+} and excessive Ba^{2+}.

$$Ca^{2+} + CO_3^{2-} =\!=\!= CaCO_3 \downarrow$$

$$4Mg^{2+} + 5CO_3^{2-} + 2H_2O =\!=\!= Mg(OH)_2 \cdot 3MgCO_3 \downarrow + 2HCO_3^-$$

$$Ba^{2+} + CO_3^{2-} =\!=\!= BaCO_3 \downarrow$$

The excessive Na_2CO_3 can be neutralized with HCl. The low content soluble impurity K^+ having a different solubility from sodium chloride and can be removed by recrystallization. It will be retained in the solution when NaCl crystals form.

【Apparatus and Chemicals】

1. Apparatus

Pallet balance; beaker; alcohol lamp; asbestos net; triangular funnel; iron stand; suction filter flask; Buchner funnel; vacuum pump; evaporating dish; measuring cylinder; crucible tongs and pH test paper.

2. Chemicals

HCl (6 mol·L^{-1}); HAc (2 mol·L^{-1}); NaOH (6 mol·L^{-1}); $BaCl_2$(0.5 mol·L^{-1}); Na_2CO_3 (saturated); $(NH_4)_2C_2O_4$ (saturated); *p*-nitroazoresorcinol (magneson); crude salt.

【Experimental Procedure】

1. Dissolving crude salt

Weigh 3 g of crude salt in a 100 mL beaker, add 20 mL of water, heat and stir to make it dissolve.

2. Removing SO_4^{2-}

Heat the solution to boiling, and then add $BaCl_2$ solution while stirring until the precipitation is complete. After continuing to boil the mixture for several minutes, vacuum filter the mixture.

3. Removing Ca^{2+}, Mg^{2+} and Ba^{2+}

Heat the above filtrate to boiling. Add saturated Na_2CO_3 solution while stirring until the precipitation is complete. Add an additional 0.5 mL of Na_2CO_3 solution, and continue heating for 5 minutes. Vacuum filter the mixture, and discard the precipitates.

4. Removing excessive CO_3^{2-}

Add 6 mol·L^{-1} HCl to the solution, heat, and stir until the pH is about 2~3.

5. Concentration and crystallization

Transfer the above solution to an evaporating dish which is already weighed. Heat and evaporate until crystals form (The volume of the solution should be about a quarter of the original solution). Cool down to room temperature, and vacuum filter the mixture. Transfer the crystals to the evaporating dish, and dry them with low heat. Cool the crystals to room temperature, weigh and yield.

6. Product purity analysis

Weigh 1 g of crude salt and purified salt, respectively. Add 5 mL of distilled water to dissolve each salt. Then perform the following qualitative analyses.

(1) SO_4^{2-} In two test tubes, transfer 1 mL of crude salt solution and 1 mL of purified salt solution, respectively. In each test tube, add two drops of 6 mol·L^{-1} HCl and two drops of 0.5 mol·L^{-1} $BaCl_2$ solution. Compare the precipitates in the two test tubes.

(2) Ca^{2+} In two test tubes, transfer 1 mL of crude salt solution and 1 mL of purified salt solution, respectively. In each test tube, add 2 mol·L^{-1} HAc, and 3~4 drops of saturated $(NH_4)_2C_2O_4$. Compare the white precipitates (CaC_2O_4) in the two test tubes.

(3) Mg^{2+} In two test tubes, transfer 1 mL of crude salt solution and 1 mL of purified salt solution, respectively. In each test tube, add five drops of 6 mol·L^{-1} NaOH and 2 drops of magneson. The blue precipitates confirm the presence of Mg^{2+}. Compare the blue precipitates in the two test tubes.

【Questions】

1. Why is the toxic $BaCl_2$ used to remove SO_4^{2-} instead of $CaCl_2$?

2. When using HCl to remove CO_3^{2-}, why should the pH be adjusted to 2~3? Can we adjust the pH to 7?

3. Can other soluble carbonates be used instead of Na_2CO_3 when removing impurity cations?

Risk Assessment and Control Worksheet

TASK	Assessor	Date
Purification of sodium chloride	Zeying WU, Hongsong WANG, Xingxing WU	1 JUL. 2021

Process Steps	Hazard Identification				Health Risk	Safety Measures				Risk Assessment				Action	
	Chemical Name or Physical Hazard	Class	Concn.	Amount	MSDS	Type of Injury Possible	To reduce or eliminate the risk	Technique	Glasses	Gloves	Fume Cbd.	Checklist~Risk Assessment		Risk Score	Risk Level	In case of Emergency/Spill	Special Comments
												Likely 1~4	Hazard 1~4		H M L		
A: Dissolving crude salt																	
Weigh 3 g of crude salt in a 100 mL beaker, add 20 mL of water, heat and stir to make it dissolve.	Alcohol burner					Burns	Handle with care	√				1	3	3	L	Report immediately to Demonstrator or Prep Room	
B: Removing SO_4^{2-}																	
Heat the solution to boiling, and then add $BaCl_2$ solution while stirring until the precipitation is complete.	Barium chloride	6,1	0.5 M	5 mL	√	Swallowing can be toxic	Handle with care		√	√		1	2	2	L	Report immediately to Demonstrator or Prep Room	
C: Removing Ca^{2+}, Mg^{2+} and Ba^{2+}																	
Heat the above filtrate to boiling. Add saturated Na_2CO_3 solution while stirring until the precipitation is complete. Add an additional 0.5 mL of Na_2CO_3 solution, and continue heating for 5 minutes. Vacuum filter the mixture, and discard the precipitates.	Saturated Na_2CO_3 solution	—		Little	√	Eye irritant	Handle with care		√	√		1	1	1	L	Sponge & water	
D: Removing excessive CO_3^{2-}																	
Add 6 M HCl to the solution, heat, and stir until the pH is about 2~3.	HCl	2,3+8	6 M	1 mL	√	Skin burns and eye injuries	Handle with care		√	√		1	3	3	L	Sponge & water	
E: Concentration and crystallization																	
F: Product purity analysis																	
Ⅰ) In two test tubes, transfer 1 mL of crude salt solution and 1 mL of purified salt solution, respectively. In each test tube, add two drops of 6 mol·L^{-1} HCl and two drops of 0.5 mol·L^{-1} $BaCl_2$ solution. Compare the precipitates in the two test tubes.																	

continued

Risk Assessment and Control Worksheet

TASK	Purification of sodium chloride				Assessor	Zeying WU, Hongsong WANG, Xingxing WU					Date	1 JUL. 2021		
Process Steps	Hazard Identification				Health Risk	Safety Measures				Checklist～Risk Assessment		Risk Assessment		Action
	Chemical Name	Class	Concn.	Amount	Type of Injury Possible	To reduce or eliminate the risk	Technique	Glasses	Gloves	Fume Cbd.	Likely	Hazard	Risk Score	In case of Emergency/Spill
		or Physical Hazard			MSDS						1~4	1~4		Risk Level
														H M L	Special Comments
Ⅱ) In two test tubes, transfer 1 mL of crude salt solution and 1 mL of purified salt solution, respectively. In each test tube, add 2 mol·L^{-1} HAc, and 3~4 drops of saturated $(NH_4)_2C_2O_4$. Compare the white precipitates (CaC_2O_4) in the two test tubes.	HAc	8+3	2 M	1 mL	√	Skin burns and eye injuries	Handle with care		√	√		1	1	1	L	Sponge & water
	Saturated $(NH_4)_2C_2O_4$ solution	—		3~4 drops	√	Eye irritant	Handle with care		√	√		1	3	3	L	Sponge & water
Ⅲ) In two test tubes, transfer 1 mL of crude salt solution and 1 mL of purified salt solution, respectively. In each test tube, add five drops of 6 M NaOH and 2 drops of magneson.	Sodium hydroxide	8	6 M	0.1 mL	√	Skin burns and eye injuries	Handle with care		√	√		1	3	3	L	Sponge & water
	Magneson	—		2 drops	√	Skin and eye irritant	Handle with care		√	√		1	1	1	L	Sponge & water

Experimental Product(s)

Experimental Waste(s)

Final assessment of overall process: Low risk when correct procedures followed

Signatures   Supervisor: Lab Manager/OHS Representative: Student/Operator:

实验 12　硫酸铜的提纯

【实验目的】

1. 了解提纯硫酸铜的原理和方法。
2. 进一步熟悉加热、溶解、蒸发、过滤、结晶等基本操作。
3. 学习用分光光度法定量检验产品中杂质铁的含量。

【实验原理】

粗硫酸铜中含有不溶性杂质和可溶性杂质 $FeSO_4$，$Fe_2(SO_4)_3$ 及其他重金属盐等。不溶性杂质可在溶解、过滤的过程中除去。粗硫酸铜中最常见的可溶杂质是 Fe^{2+} 和 Fe^{3+}。为了去除铁离子，首先将 Fe^{2+} 用氧化剂 H_2O_2 或 Br_2 氧化成 Fe^{3+}，然后调节溶液的 pH，使 Fe^{3+} 水解为 $Fe(OH)_3$ 沉淀而除去，反应的离子议程式如下：

$$2Fe^{2+} + H_2O_2 + 2H^+ =\!=\!= 2Fe^{3+} + 2H_2O$$

$$Fe^{3+} + 3H_2O =\!=\!= Fe(OH)_3\downarrow + 3H^+ \,(pH \approx 4.0)$$

除铁离子后的滤液，经蒸发，即得晶体，其他微量的可溶性杂质仍留在母液中，通过减压抽滤可与硫酸铜晶体分离。

【器材与试剂】

1. 器材：天平，酒精灯，721 型分光光度计，玻璃漏斗，布氏漏斗，抽滤瓶，蒸发皿，25 mL 比色管，量筒，研钵。
2. 试剂：粗 $CuSO_4$，H_2SO_4（2 mol·L^{-1}），HCl 溶液（2 mol·L^{-1}），H_2O_2（质量分数为 3%），NaOH（2 mol·L^{-1}），$NH_3·H_2O$（6 mol·L^{-1}），KSCN（1 mol·L^{-1}），滤纸，pH 试纸。

【实验内容】

1. 粗硫酸铜的提纯

称取研细的粗 $CuSO_4$ 2 g 于 250 mL 烧杯中，加入 8 mL 去离子水，加热并不断搅拌，使其溶解。随后，在边加热边搅拌下，在溶液中滴加 1 mL 3% H_2O_2 溶液，继续逐滴滴加 2 mol·L^{-1} NaOH 溶液直至溶液 pH≈4，加热并煮沸数分钟后趁热常压过滤，将滤液转移至蒸发皿中。

用 2 mol·L^{-1} H_2SO_4 溶液调节滤液 pH 至 1~2，加热、蒸发浓缩至溶液表面刚出现一层结晶膜时，停止加热，冷却至室温。将析出的晶体用布氏漏斗减压抽滤至干，取出晶体，用滤纸吸干其表面水分，称量，计算产率。

2. 产品纯度的检验

称取 0.2 g 提纯后的硫酸铜晶体,放入小烧杯中,用 3 mL 去离子水溶解,加 2 滴 2 mol·L^{-1} H$_2$SO$_4$ 溶液酸化,再加入 10 滴 3% H$_2$O$_2$ 溶液,充分搅拌,煮沸片刻,将溶液中的 Fe^{2+} 全部氧化为 Fe^{3+}。待溶液冷却后,逐滴加入 6 mol·L^{-1} 氨水,不断搅拌直至生成的浅蓝色 Cu$_2$(OH)$_3$SO$_4$ 沉淀溶解为深蓝色 [Cu(NH$_3$)$_4$]$^{2+}$ 溶液为止。用抽滤瓶过滤,并用去离子水洗去滤纸上的蓝色,弃去滤液。

如有 Fe(OH)$_3$ 沉淀留在滤纸上,用 1.5 mL(约 30 滴)热的 2 mol·L^{-1} HCl 溶液溶解滤纸上的 Fe(OH)$_3$ 沉淀,并将抽滤瓶洗净接收滤液。可重复上述操作,直到 Fe(OH)$_3$ 全部溶解。

在滤液中加入 2 滴 1 mol·L^{-1} KSCN 溶液,并用去离子水稀释至 5 mL,摇匀。

将上述溶液倒入 1 cm 比色皿中(液面高度不要超过比色皿 3/4 高度),以去离子水为参比液,用 721 型分光光度计在波长为 465 nm 处测定其吸光度(A)。在 $A-w$(Fe^{3+})标准曲线上查出与 A 对应的 Fe^{3+} 的质量分数 w,与表 12-1 中产品规格对照,可确定产品的规格。

表 12-1 产品规格

规格	分析纯	化学纯
w(Fe^{3+})×100	0.003	0.02

【思考题】

1. 粗硫酸铜中的杂质 Fe^{2+} 为什么要氧化成 Fe^{3+} 除去?采用 H$_2$O$_2$ 作氧化剂比其他氧化剂有什么优点?

2. 滤液蒸发前为什么将其 pH 调至 1～2?

3. 用 KSCN 检验 Fe^{3+} 时为什么要加盐酸?

• 视频:CuSO$_4$·5H$_2$O 制备

风险评估及控制工作表

实验任务	硫酸铜的提纯					评估人	吴泽颖、王红松、吴星星				评估日期	2021年7月1日			
过程步骤	危害鉴别					健康风险评估		安全措施			清单/风险评估	风险评估			措施
	化合物名称	类别	浓度	用量	化学品安全技术说明书	可能的伤害种类	降低或消除风险	技术	防护眼镜	手套	通风橱	可能性 1~4	危害 1~4	风险	危害级别 高中低	紧急情况/液溅 特殊备注
		或物理危害														
A:粗硫酸铜的提纯																
Ⅰ) 称取研细的粗 CuSO₄ 2 g 于 250 mL 烧杯中,加入 8 mL 去离子水,加热并不断搅拌,使其溶解。随后,在边加热边搅拌下,在溶液中滴加 1 mL 3% H₂O₂ 溶液,继续逐滴滴加 2 mol·L⁻¹ NaOH 溶液直至溶液 pH≈4,加热并煮沸数分钟后趁热常压过滤,将滤液转移至蒸发皿中。	酒精灯					烧伤	关注	√				1	3	3	低	立即向指导老师或实验员报告
	粗 CuSO₄	—		2 g	√	皮肤刺激	关注		√			1	2	2	低	大量水冲洗
	H₂O₂	5.1+8	3%	1 mL	√	皮肤灼伤和眼损伤	关注		√	√		1	3	3	低	大量水冲洗
	NaOH	8	2 M	20 mL	√	皮肤灼伤和眼损伤	关注		√	√		1	3	3	低	大量水冲洗
Ⅱ) 用 2 mol·L⁻¹ H₂SO₄ 溶液调节滤液 pH 至 1~2,加热,蒸发浓缩至溶液表面刚出现一层结晶膜时,停止加热,冷却至室温。将析出的晶体用布氏漏斗减压抽滤至干,取出晶体,用滤纸吸干其表面水分,称量,计算产率。	H₂SO₄	8	2 M	10 mL	√	皮肤灼伤和眼损伤	关注		√	√		1	2	2	低	大量水冲洗

续表

风险评估及控制工作表

实验任务：硫酸铜的提纯
评估人：吴泽颖、王红松、吴星星
评估日期：2021 年 7 月 1 日

化合物名称	危害鉴别			化学品安全技术说明书	健康风险		安全措施			风险评估清单	风险评估（1~4）			危害级别（高中低）	措施	
	类别	浓度	用量		可能的伤害种类	降低或消除风险	技术	防护眼镜	手套	通风橱		可能性	危害	风险		紧急情况/泼洒	特殊备注
	或物理危害																

B: 产品纯度的检验

实验步骤：

Ⅰ）称取 0.2 g 提纯后的硫酸铜晶体，放入小烧杯中，用 3 mL 去离子水溶解，加 2 滴 2 mol·L⁻¹ H_2SO_4 溶液酸化，再加入 10 滴 3% H_2O_2 溶液，充分搅拌，煮沸片刻，将溶液中的 Fe^{2+} 全部氧化为 Fe^{3+}。待溶液冷却后，逐滴加入 6 mol·L⁻¹ 氨水，不断搅拌直至生成的浅蓝色 $Cu_2(OH)_2SO_4$ 沉淀溶解为深蓝色 $[Cu(NH_3)_4]^{2+}$ 溶液为止。用抽滤瓶过滤，并用去离子水洗去滤纸上的蓝色去滤液。

Ⅱ）如有 $Fe(OH)_3$ 沉淀留在滤纸上，用 1.5 mL 2 mol·L⁻¹ HCl 溶液溶解滤纸上的 $Fe(OH)_3$ 沉淀，并将抽滤瓶洗净接收滤液。可重复上述操作，直到 $Fe(OH)_3$ 全溶解。

Ⅲ）在滤液中加入 2 滴 1 mol·L⁻¹ KSCN 溶液，并用去离子水稀释至 5 mL，摇匀。

化合物	类别	浓度	用量	化学品安全技术说明书	可能的伤害	降低或消除风险	技术	防护眼镜	手套	通风橱	可能性	危害	风险	危害级别	紧急情况/泼洒
H_2SO_4	8	2 M	2 滴	√	皮肤灼伤和眼损伤	关注		√	√	√	1	2	2	低	大量水冲洗
H_2O_2	5,1+8	3%	10 滴	√	皮肤灼伤和眼损伤	关注		√	√	√	1	3	3	低	大量水冲洗
氨水	8	6 M	10 mL	√	皮肤灼伤和眼损伤	关注		√	√	√	1	3	3	低	大量水冲洗
HCl	2,3+8	2 M	1.5 mL	√	皮肤灼伤和眼损伤	关注		√	√	√	1	4	4	低	大量水冲洗
KSCN	—	1 M	2 滴	√	眼损伤，皮肤接触有害	关注		√	√	√	1	1	1	低	大量水冲洗

续表

风险评估及控制工作表

实验任务	硫酸铜的提纯			评估人	吴泽颖、王红松、吴星星			评估日期	2021年7月1日		

过程步骤	危害鉴别			健康风险	降低或消除风险	安全措施			清单/风险评估	风险评估			措施
	化合物名称	类别 浓度 用量	化学品安全技术说明书	可能的伤害种类		技术	防护眼镜	手套	通风橱	可能性	危害	风险	危害级别	紧急情况/泼洒特殊备注
		或物理危害								1~4	1~4		高中低	
Ⅳ）将上述溶液倒入1 cm比色皿中（液面高度不要超过比色皿3/4高度），以去离子水为参比液，用721型分光光度计在波长为465 nm处测定其吸光度（A）。在A—w对应Fe^{3+}标准曲线上查出与A对应的质量分数w，与表12-1中产品规格对照，可确定产品的规格。	分光光度计			电击	确保电线已贴上标签		√	√		1	3	3	低	向实验老师或实验备员报告。
实验产物														
实验废弃物														
整个过程的最终评估	低风险（按照正确流程操作）													
签名	实验老师：				实验室主任/职业安全健康（OHS）代表：					学生/操作人：				

Experiment 12 Purification of copper (II) sulfate

【Objectives】

1. To understand the principle and method of purifying copper sulfate.
2. To practice heating, dissolution, evaporation, filtration, crystallization and other basic operations.
3. To learn to quantitatively test the content of impurity iron in products by spectrophotometry.

【Principles】

Crude copper sulfate is a mixture of copper (II) sulfate with various impurities that may include iron (II) sulfate ($FeSO_4$), iron (III) sulfate [$Fe_2(SO_4)_3$], and other metal salts. Insoluble impurities can be removed by dissolution and subsequent filtration. The most common soluble impurities in crude copper sulfate are iron (II) sulfate and iron (III) sulfate. To remove iron ions, Fe^{2+} is oxidized to Fe^{3+} first by using H_2O_2 or Br_2. Then Fe^{3+} will be removed as a $Fe(OH)_3$ precipitate by adjusting the pH of the solution, as shown in the following reactions:

$$2Fe^{2+} + H_2O_2 + 2H^+ = 2Fe^{3+} + 2H_2O$$

$$Fe^{3+} + 3H_2O = Fe(OH)_3 \downarrow + 3H^+ (pH \approx 4.0)$$

The filtrate without Fe ions can be recrystallized. When copper sulfate crystals form in the solution, some other trace amount of soluble impurities, which have not reached saturation, will remain in solution. Therefore pure copper sulfate can be separated by filtration.

【Apparatus and Chemicals】

1. Apparatus

Electronic balance; alcohol lamp; 721 spectrophotometer; funnel; Buchner funnel; filter flask; evaporating dish; 25 mL colorimetric tube; volumetric cylinder; mortar.

2. Chemicals

Crude $CuSO_4$; H_2SO_4 (2 mol·L^{-1}); HCl (2 mol·L^{-1}, AR); H_2O_2 (3%, wt/wt); NaOH (2 mol·L^{-1}); $NH_3·H_2O$ (6 mol·L^{-1}); KSCN (1 mol·L^{-1}); filter paper; pH test paper.

【Experimental Procedures】

1. Purification of crude copper sulfate

Weigh 2 g of a ground powder sample of crude $CuSO_4$, and transfer it into a 250 mL beaker. Add

8 mL of distilled water to the beaker, heat the mixture, and stir using a stirring rod until it dissolves. Add 1 mL of 3% H_2O_2 to the above solution. Use 2 mol·L^{-1} NaOH to adjust the pH of the solution to about 4. Heat and boil the solution for a few minutes. Then filter the solution. Transfer the filtrate to an evaporating dish.

Add 2 mol·L^{-1} H_2SO_4 to adjust the pH value of solution to 1~2. Heat to evaporate the solution until crystals begin to form a thin layer on the surface. Cool the solution to room temperature. Precipitated crystals were filtered to dryness by means of a Buchner funnel under reduced pressure. The crystals were taken out, and the surface water was sucked dry with filter paper. Finally, weigh the crystals and calculate the yield.

2. Inspection of product purity

Weigh 0.2 g purified copper sulfate crystal, put it into a small beaker, dissolve it with 3 mL deionized water, add 2 drops of 2 mol·L^{-1} H_2SO_4 solution for acidification, then add 10 drops of 3% H_2O_2 solution, stir it fully, boil it for a while, and oxidize all Fe^{2+} in the solution to Fe^{3+}. After the solution is cooled, add 6 mol·L^{-1} ammonia drop by drop and stir continuously until the formed light blue $Cu_2(OH)_3SO_4$ precipitation dissolves into dark blue $[Cu(NH_3)_4]^{2+}$ solution. Filter it with suction bottle and wash the filter paper with deionized water till the paper is colorless, then discard the filtrate.

If there is $Fe(OH)_3$ precipitate on the filter paper, drop 1.5 mL (about 30 drops) of hot 2 mol·L^{-1} HCl solution onto the filter paper with a dropper to dissolve the $Fe(OH)_3$ precipitate, and wash the filter bottle to receive the filtrate. The above operation can be repeated until all $Fe(OH)_3$ precipitate is dissolved.

Add 2 drops of 1 mol·L^{-1} KSCN solution to the filtrate, dilute to 5 mL with deionized water, and shake well.

Pour the above solution into a 1 cm cuvette (the liquid level should not exceed 3/4 of the cuvette height), use deionized water as the reference solution, and use 721 spectrophotometer to measure the absorbance (A) at 465 nm. Find out the mass fraction w of Fe^{3+} corresponding to A on the $A-w$(Fe^{3+}) standard curve, and compare it with the product specifications in Table 12-1 to determine the product specifications.

Table 12-1 Product specifications

specifications	AR	CP
$w(Fe^{3+}) \times 100$	0.003	0.02

【Questions】

1. Why should impurity Fe^{2+} be oxidized to Fe^{3+} before removal? What is the advantage of using H_2O_2 as an oxidant?

2. Why adjust the pH of filtrate to 1~2 before evaporation?

3. Why should hydrochloric acid be added when testing Fe^{3+} with KSCN?

Risk Assessment and Control Worksheet

TASK	Purification of copper (Ⅱ) sulfate					Assessor	Zeying WU, Hongsong WANG, Xingxing WU				Date	1 JUL. 2021			
Process Steps	Hazard Identification					Health Risk	Safety Measures				Checklist～Risk Assessment			Action	
	Chemical Name	Class	Concn.	Amount	MSDS	Type of Injury Possible	To reduce or eliminate the risk	Technique	Glasses	Gloves	Fume Cbd.	Likely 1~4	Hazard 1~4	Risk Score	Risk Level H M L	In case of Emergency/Spill	Special Comments
		or Physical Hazard														
	alcohol burner					Burns	Handle with care	√				1	3	3	L	Report immediately to Demonstrator or Prep Room
	Crude CuSO$_4$	—		2 g	√	Skin irritant	Handle with care		√	√		1	2	2	L	Sponge & water
	H$_2$O$_2$	5.1+8	3%	1 mL	√	Skin burns and eye injuries	Handle with care		√	√		1	3	3	L	Sponge & water
	NaOH	8	2 M	20 mL	√	Skin burns and eye injuries	Handle with care		√	√		1	3	3	L	Sponge & water
	H$_2$SO$_4$	8	2 M	10 mL	√	Skin burns and eye injuries	Handle with care		√	√		1	2	2	L	Sponge & water

A: Purification of crude copper sulfate

Ⅰ) Weigh 2 g of a ground powder sample of crude CuSO$_4$, and transfer it into a 250 mL beaker. Add 8 mL of distilled water to the beaker, heat the mixture, and stir using a stirring rod until it dissolves. Add 1 mL of 3% H$_2$O$_2$ to the above solution. Use 2 mol·L^{-1} NaOH to adjust the pH of the solution to about 4. Heat and boil the solution for a few minutes. Then filter the solution. Transfer the filtrate to an evaporating dish.

Ⅱ) Add 2 mol·L^{-1} H$_2$SO$_4$ to adjust the pH value of solution to 1～2. Heat to evaporate the solution until crystals begin to form a thin layer on the surface. Cool the solution to room temperature. Precipitated crystals were filtered to dryness by means of a Buchner funnel under reduced pressure. The crystals were taken out, and the surface water was sucked dry with filter paper. Finally, weigh the crystals and calculate the yield.

continued

Risk Assessment and Control Worksheet

TASK	Purification of copper (II) sulfate				Assessor	Zeying WU, Hongsong WANG, Xingxing WU				Date	1 JUL. 2021		
Process Steps	Hazard Identification				Health Risk	Safety Measures				Checklist↘Risk Assessment	Risk Assessment		Action
	Chemical Name	Class	Concn.	Amount	MSDS	Type of Injury Possible	To reduce or eliminate the risk	Technique	Glasses	Gloves	Fume Cbd.	Likely	Hazard	Risk Score	Risk Level	In case of Emergency/Spill
		or Physical Hazard										$1\sim4$	$1\sim4$		H M L	Special Comments

B: Inspection of product purity

Process Steps	Chemical Name	Class	Concn.	Amount	MSDS	Type of Injury Possible	To reduce or eliminate the risk	Technique	Glasses	Gloves	Fume Cbd.	Likely	Hazard	Risk Score	Risk Level	In case of Emergency/Spill
I) Weigh 0.2 g purified copper sulfate crystal, put it into a small beaker, dissolve it with 3 mL deionized water, add 2 drops of $2\ mol \cdot L^{-1}\ H_2SO_4$ solution for acidification, then add 10 drops of $3\%\ H_2O_2$ solution, stir it fully, boil it for a while, and oxidize all Fe^{3+} in the solution to Fe^{3+}. After the solution is cooled, add 6 mol·L^{-1} ammonia drop by drop and stir continuously until the formed light blue $Cu_2(OH)_3SO_4$ precipitation dissolves into dark blue $[Cu(NH_3)_4]^{2+}$ solution. Filter it with suction bottle and wash the filter paper with deionized water till the paper is colorless, then discard the filtrate.	H_2SO_4	8	2 M	2 drops	√	Skin burns and eye injuries	Handle with care		√	√		1	2	2	L	Sponge & water
	H_2O_2	5.1+8	3%	10 drops	√	Skin burns and eye injuries	Handle with care		√	√		1	3	3	L	Sponge & water
	$NH_3 \cdot H_2O$	8	6 M	10 mL	√	Skin burns and eye injuries	Handle with care		√	√	√	1	3	3	L	Sponge & water

continued

Risk Assessment and Control Worksheet

TASK	Assessor	Date
Purification of copper (II) sulfate	Zeying WU, Hongsong WANG, Xingxing WU	1 JUL. 2021

Process Steps	Hazard Identification				Health Risk	Safety Measures				Risk Assessment			Action	
	Chemical Name	Class	Concn.	Amount	MSDS	Type of Injury Possible	To reduce or eliminate the risk	Technique	Glasses	Gloves	Fume Cbd.	Checklist＼Risk Assessment			In case of Emergency/Spill	Special Comments

Process Steps	Chemical Name	Class	Concn.	Amount	MSDS	Type of Injury Possible	To reduce or eliminate the risk	Technique	Glasses	Gloves	Fume Cbd.	Likely 1~4	Hazard 1~4	Risk Score	Risk Level H M L	In case of Emergency/Spill	Special Comments
II) If there is Fe(OH)$_3$ precipitate on the filter paper, drop 1.5 mL (about 30 drops) of hot 2 mol·L^{-1} HCl solution onto the filter paper with a dropper to dissolve the Fe(OH)$_3$ precipitate, and wash the filter bottle to receive the filtrate. The above operation can be repeated until all Fe(OH)$_3$ precipitate is dissolved.	HCl	2.3+8	2 M	1.5 mL	√	Skin burns and eye injuries	Handle with care		√	√		1	4	4	L	Sponge & water	
III) Add 2 drops of 1 mol·L^{-1} KSCN solution to the filtrate, dilute to 5 mL with deionized water, and shake well.	KSCN	—	1 M	2 drops	√	Skin burns and eye injuries	Handle with care		√	√	√	1	1	1	L	Sponge & water	

continued

Risk Assessment and Control Worksheet

TASK	Purification of copper (Ⅱ) sulfate				Assessor	Zeying WU, Hongsong WANG, Xingxing WU				Date	1 JUL. 2021		
Process Steps	Hazard Identification				Health Risk	Safety Measures				Risk Assessment			Action
	Chemical Name	Class	Concn.	Amount	MSDS	Type of Injury Possible	To reduce or eliminate the risk	Technique	Glasses	Gloves	Fume Cbd.	Checklist～Risk Assessment	In case of Emergency/Spill
		or Physical Hazard										Likely	Hazard	Risk Score	Risk Level	Special Comments
												$1\sim4$	$1\sim4$		H M L	
Ⅳ) Pour the above solution into a 1 cm cuvette (the liquid level should not exceed 3/4 of the cuvette height), use deionized water as the reference solution, and use 721 spectrophotometer to measure the absorbance (A) at 465 nm. Find out the mass fraction w of Fe^{3+} corresponding to A on the $A-w$ (Fe^{3+}) standard curve, and compare it with the product specifications in table 12.1 to determine the product specifications.	Spectrophotometer					Electric shock	Ensure cord is tagged	✓	✓			1	3	3	L	Report to Demonstrator or Prep Room Staff
Experimental Product (s)																
Experimental Waste (s)																
Final assessment of overall process					Low risk when correct procedures followed											
Signatures	Supervisor:				Lab Manager/OHS Representative:							Student/Operator:				

实验 13　NaOH 标准溶液的配制与标定及铵盐中氮含量的测定

【实验目的】

1. 学会配制标准溶液和利用基准物质对其浓度进行标定的方法。
2. 基本掌握滴定操作和滴定终点的判断。
3. 掌握甲醛法测定铵盐中氮含量的原理及操作方法。
4. 掌握酸碱指示剂的选择原理。

【实验原理】

1. NaOH 标准溶液的配制与标定

NaOH 标准溶液采用间接配制法配制,采用基准物质进行标定。标定 NaOH 标准溶液的常用基准物有:邻苯二甲酸氢钾(KHP)、草酸。本实验采用 KHP 作为基准物质标定 NaOH 标准溶液。其标定反应为:

$$KHP + NaOH = KNaP + H_2O$$

反应产物为二元弱碱,在溶液中显弱碱性,可选用酚酞作指示剂。

滴定终点颜色变化:由无色变为微红色,且半分钟不褪色。

2. 铵盐中氮含量的测定

NH_4Cl、$(NH_4)_2SO_4$ 等铵盐是常用的氮肥,肥效的高低主要取决于其含氮量,即以 NH_4^+ 形式存在的氮含量。由于铵盐中 NH_4^+ 酸性太弱,其 K_a 值为 5.6×10^{-10},$c \cdot K_a < 10^{-8}$,不能用 NaOH 标准溶液直接滴定,因此,在实际生产和实验室中常采用甲醛法使弱酸强化以测定铵盐中的含氮量。反应按下式进行:

$$4NH_4^+ + 6HCHO = (CH_2)_6N_4H^+ + 3H^+ + 6H_2O$$

$$(CH_2)_6N_4H^+ + 3H^+ + 4OH^- = (CH_2)_6N_4 + 4H_2O$$

被甲醛强化生成的 $(CH_2)_6N_4H^+$ 的 K_a 为 7.1×10^{-6},$c \cdot K_a > 10^{-8}$,可直接用 NaOH 标准溶液滴定。

【器材与试剂】

1. 器材:25 mL 碱式滴定管,250 mL 锥形瓶,250 mL 容量瓶,1 000 mL 容量瓶,250 mL 烧杯,10 mL 移液管,25 mL 移液管,电子天平。
2. 试剂:NaOH,NH_4Cl,$(NH_4)_2SO_4$,邻苯二甲酸氢钾(KHP),酚酞指示剂($2 \text{ g} \cdot \text{L}^{-1}$ 乙醇溶液),

40%甲醛。

【实验内容】

1. NaOH 饱和溶液的配制

称取 120 g NaOH 于 250 mL 烧杯中,加入 100 mL 水,搅拌溶解。

2. 0.1 mol·L^{-1} NaOH 溶液的配制

量取 5.6 mL NaOH 饱和溶液的上清液,转入 1 000 mL 容量瓶中,加水稀释,定容,充分摇匀。

3. 0.1 mol·L^{-1} NaOH 溶液的标定

准确称取 0.4~0.5 g 已烘干至恒重的邻苯二甲酸氢钾于 250 mL 锥形瓶中,加入 20~30 mL 水,温热使之溶解,冷却后加 1~2 滴酚酞指示剂,用 0.1 mol·L^{-1} NaOH 溶液滴定至溶液由无色转变为微红色,且半分钟不褪色,记录所消耗的 NaOH 溶液的体积,平行标定 3 份,计算 NaOH 标准溶液浓度。将实验数据记录于表 13-1。

表 13-1 NaOH 溶液的标定

项目 \ 序号	1	2	3
m_{KHP}＋称量瓶(倾出前)/g			
m_{KHP}＋称量瓶(倾出后)/g			
m_{KHP}/g			
V_{NaOH}终读数/mL			
V_{NaOH}初读数/mL			
V_{NaOH}/mL			
$c_{NaOH}/(mol·L^{-1})=(m/M)_{KHP}/V_{NaOH}$			
$\bar{c}_{NaOH}/(mol·L^{-1})$			
相对平均偏差/%			

4. 铵盐样品的准备

准确称取铵盐试样 1.6~1.7 g 于小烧杯中,用适量蒸馏水溶解后全部转移至 250 mL 容量瓶中,用蒸馏水稀释至刻度,摇匀。用移液管移取 20.00 mL 试液于锥形瓶中。

因甲醛中常含有微量甲酸,为消除其对滴定结果的影响,应予以去除,具体方法:以酚酞为指示剂,用 0.1 mol·L^{-1} NaOH 溶液中和至溶液呈淡红色。

向试液中加入 5 mL 上述甲醛溶液、1~2 滴酚酞指示剂,摇匀,静置 1 min。

5. 滴定

用 NaOH 标准溶液滴定至溶液由无色转变为淡红色,且半分钟不褪色,记录消耗 NaOH 标准溶液体积 V(mL),平行测定 3 次。计算试样中的含氮量及相对平均偏差。

【注意事项】

1. 用来配制 NaOH 溶液的水应在使用前加热煮沸,放冷后再使用,以除去其中的 CO_2。
2. 配制 NaOH 溶液时,一定要待 Na_2CO_3 沉淀后再取 NaOH 饱和溶液的上层清液。
3. 如果铵盐中含有游离酸,也应将其去除,具体方法:先加甲基红指示剂,用 NaOH 溶液滴定至溶液呈橙色,然后再加入甲醛溶液进行测定。

【思考题】

1. 如何计算所需基准物邻苯二甲酸氢钾的质量范围?称得太多或太少对标定结果有何影响?
2. 溶解基准物质时加入 20~30 mL 水,是用量筒量取,还是用移液管移取?为什么?
3. 如果基准物未烘干,将使标准溶液浓度的标定结果偏高还是偏低?
4. 以酚酞作指示剂,用 NaOH 标准溶液标定 HCl 溶液浓度时,若 NaOH 溶液因贮存不当吸收了 CO_2,对测定结果有何影响?
5. 铵盐中氮含量的测定为何不采用 NaOH 直接滴定法?
6. 为什么中和甲醛试剂中的甲酸以酚酞作指示剂,而中和铵盐试样中的游离酸则以甲基红作指示剂?
7. NH_4HCO_3 中含氮量的测定,能否用甲醛法?

● 酸碱标定

风险评估及控制工作表

实验任务	NaOH 标准溶液的配制与标定及铵盐中氮含量的测定				评估人	吴泽颖,王红松,吴星星				评估日期	2021 年 7 月 1 日		
过程步骤	危害鉴别				健康风险	降低或消除风险	安全措施			清单/风险评估	风险评估		紧急情况/泄洒措施
	化合物名称	类别	浓度	用量	化学品安全技术说明书	可能的伤害种类		技术	防护眼镜	手套	通风橱		可能性	危害	风险	危害级别	特殊备注
		或物理危害											1~4	1~4		高中低	

A: NaOH 饱和溶液的配制

过程步骤	化合物名称	类别	浓度	用量	化学品安全技术说明书	可能的伤害种类	降低或消除风险	技术	防护眼镜	手套	通风橱	可能性	危害	风险	危害级别	紧急情况/泄洒措施
称取120 g NaOH 于250 mL 烧杯中,加入100 mL水,搅拌溶解。	NaOH	8			√	皮肤灼伤和眼损伤	关注		√	√		1	3	3	低	大量水冲洗

B: 0.1 mol·L⁻¹ NaOH 溶液的配制

过程步骤	化合物名称	类别	浓度	用量	化学品安全技术说明书	可能的伤害种类	降低或消除风险	技术	防护眼镜	手套	通风橱	可能性	危害	风险	危害级别	紧急情况/泄洒措施
量取5.6 mL NaOH 饱和溶液的上清液,转入1 000 mL 容量瓶中,加水稀释,定容,充分摇匀。	NaOH	8	0.1 M	1 000 mL	√	皮肤灼伤和眼损伤	关注		√	√		1	3	3	低	大量水冲洗

C: 0.1 mol·L⁻¹ NaOH 溶液的标定

过程步骤	化合物名称	类别	浓度	用量	化学品安全技术说明书	可能的伤害种类	降低或消除风险	技术	防护眼镜	手套	通风橱	可能性	危害	风险	危害级别	紧急情况/泄洒措施
准确称取 0.4~0.5 g 已烘干的邻苯二甲酸氢钾于250 mL 锥形瓶中,加入20~30 mL水,温热使之溶解,冷却后加1~2滴酚酞指示剂,用0.1 mol·L⁻¹ NaOH 溶液滴定至溶液由无色转变为微红色,且半分钟不褪色,记录所消耗的 NaOH 溶液的体积,平行标定3份,计算 NaOH 标准溶液浓度。	KHP	5		0.5 g	√	吞咽有害	关注		√	√		1	3	3	低	向实验老师或准备实验员报告
	NaOH	8	0.1 M	1 000 mL	√	皮肤灼伤和眼损伤	关注		√	√		1	3	3	低	大量水冲洗
	酚酞指示剂	—	1%	1 mL	—	—			√	√		1	1	1	低	大量水冲洗

D: 铵盐样品的准备

过程步骤	化合物名称	类别	浓度	用量	化学品安全技术说明书	可能的伤害种类	降低或消除风险	技术	防护眼镜	手套	通风橱	可能性	危害	风险	危害级别	紧急情况/泄洒措施
1) 准确称取铵盐试样1.6~1.7 g于小烧杯中,用适量蒸馏水溶解后全部转移至250 mL 容量瓶中,用蒸馏水稀释至刻度,摇匀。用移液管移取20.00 mL 试液于锥形瓶中。	NH_4Cl	—	0.1 M	250 mL	√	眼刺激	关注		√	√		1	2	2	低	大量水冲洗
	$(NH_4)_2SO_4$	—	0.1 M	250 mL	√	—	关注		√	√		1	2	2	低	大量水冲洗

续表

风险评估及控制工作表

实验任务	NaOH 标准溶液的配制与标定及铵盐中氮含量的测定				评估人	吴泽颖、王红松、吴星星				评估日期	2021年7月1日			
过程步骤	危害鉴别				健康风险		安全措施			清单/风险评估	风险评估			措施
	化合物名称	类别	浓度	用量	化学品安全技术说明书	可能的伤害种类	降低或消除风险	技术	防护眼镜	手套	通风橱		可能性	危害	风险	危害级别	紧急情况/泼洒特殊备注
	或物理危害															
II）因甲醛中常有微量甲酸，为消除其对滴定结果的影响，应予以去除，具体方法：以酚酞为指示剂，用 0.1 mol·L⁻¹ NaOH 溶液中和至溶液淡呈红色。向试液中加入 5 mL 上述甲醛溶液，1~2 滴酚酞指示剂，摇匀，静置 1 min。	HCHO	8	40%	20 mL	√	皮肤灼伤和眼损伤	关注		√	√	√		1	4	4	低	大量水冲洗
E：滴定													1~4	1~4		高中低	
实验产物																	
实验废弃物																	
整个过程的最终评估	低风险（按照正确流程操作）																
签名	实验老师：					实验室主任/职业安全健康（OHS）代表：							学生/操作人：				

Experiment 13 Preparation and standardization of NaOH standard solution and determination of nitrogen content of ammonium salt

【Objectives】

1. To learn to prepare standard solution and standardize concentration using standard regent.
2. To master procedures of titration and judgement of endpoint.
3. To master the determination principle of nitrogen content of ammonium salt using formaldehyde method.
4. To master the principle of selecting acid-base indicator.

【Principles】

1. Preparation and standardization of NaOH solution

NaOH solution is prepared by indirect preparation method and standardized by primary standard. The commonly used standards for standardizing NaOH solution are potassium hydrogen phthalate (KHP) and oxalic acid. In this experiment, KHP was used as primary standard to standardize NaOH solution. The titration reaction is as follows:

$$KHP + NaOH = KNaP + H_2O$$

The reaction product KHP is a binary weak base, which is weakly alkaline in solution. Phenolphthalein can be used as indicator.

Color change of titration end point: colorless ⟶ reddish (the color stands for 30 s)

2. Determination of nitrogen content of ammonium salt

NH_4Cl, $(NH_4)_2SO_4$ and other ammonium salts are commonly used as nitrogenous fertilizers. For K_a of NH_4^+ is 5.6×10^{-10}, and $c \cdot K_a < 10^{-8}$, the acidity of ammonium salt is too weak to titrate directly using NaOH standard solution. Therefore, formaldehyde method is used widely to determine the nitrogen content of ammonium salt to strengthen weak acid in production experiment. The titration reaction is below:

$$4NH_4^+ + 6HCHO = (CH_2)_6N_4H^+ + 3H^+ + 6H_2O$$

$$(CH_2)_6N_4H^+ + 3H^+ + 4OH^- = (CH_2)_6N_4 + 4H_2O$$

The K_a of $(CH_2)_6N_4H^+$ which strengthened by formaldehyde is 7.1×10^{-6}, $c \cdot K_a > 10^{-8}$, which can be directly titrated with NaOH standard solution.

【Apparatus and Chemicals】

1. Apparatus

Alkali burette (25 mL); Erlenmeyer flask (250 mL); Volumetric flask (250 mL); Volumetric flask (1 000 mL); Beaker (250 mL); Pipette (10 mL); Pipette (25 mL); Electronic balance.

2. Chemicals

NaOH; NH_4Cl; $(NH_4)_2SO_4$; potassium hydrogen phthalate (KHP); phenolphthalein indicator (2 g·L^{-1} ethanol solution); 40% formaldehyde.

【Experimental Procedures】

1. Preparation of NaOH saturated solution

Dissolve 120 g of NaOH in 100 mL of water and mix well.

2. Preparation of NaOH solution (0.1 mol·L^{-1})

Transfer 5.6 mL of the supernatant of NaOH saturated solution to a volumetric flask (1 000 mL) and dilute to the mark with water.

3. Standardization of NaOH solution (0.1 mol·L^{-1})

Accurately weigh 0.4~0.5 g of pre-dried KHP into an Erlenmeyer flask (250 mL), add 20~30 mL of water. Heat the flask to dissolve and add 1~2 drops of phenolphthalein indicator. Titrate with 0.1 mol·L^{-1} NaOH standard solution until the color of the solution turns from colorless to reddish and the color stands for 30 s. Record the volume of the NaOH standard solution consumed in the titration. Repeat 3 times. Calculate the concentration of the NaOH standard solution. Record the experimental data in table 13-1.

Table 13-1 Standardization of NaOH solution

Data type \ Sample	1	2	3
m_{KHP}+weighing bottle (before pouring out) /g			
m_{KHP}+weighing bottle (after pouring out) /g			
m_{KHP}/g			
V_{NaOH}(final reading) /mL			
V_{NaOH}(Initial reading) /mL			
V_{NaOH}/mL			
c_{NaOH}/(mol·L^{-1})=$(m/M)_{KHP}/V_{NaOH}$			
$\bar{c}_{NaOH}$/(mol·L^{-1})			
RAD/%			

4. Preparation of ammonium salt sample

Accurately weigh 1.6~1.7 g of ammonium salt, dissolve, transfer to a volumetric flask (250 mL), dilute to the mark with distilled water and shake well. Then transfer 20.00 mL of ammonium salt solution to an Erlenmeyer flask (250 mL).

A little of formic acid, contained in formaldehyde solution, can be neutralized by 0.1 mol·L^{-1} NaOH solution using phenolphthalein as the indicator. When the solution is reddish, free acid is removed.

Add 5 mL of above formaldehyde solution and 1~2 drops of phenolphthalein to the ammonium salt solution, shake well and stand for 1 min.

5. Titration

Titrate the solution with NaOH standard solution until it is reddish and not fades within 30 s. Record the volume of NaOH standard solution consumed. Repeat 3 times. Calculate the nitrogen content of ammonium salt and relative average deviation.

【Notes】

1. The freshly boiled and cooled distilled water should be used to prepare NaOH solution to remove $CO_2(g)$.

2. When preparing NaOH solution, the supernatant of NaOH saturated solution should be obtained after precipitation of Na_2CO_3.

3. If ammonium salt solution contains free acids, it should be neutralized and removed by NaOH solution until the solution is orange using methyl red as the indicator before determination.

【Questions】

1. How to calculate the mass range of potassium hydrogen phthalate? How does too much or too little potassium hydrogen affect the standardization results?

2. When 20~30 mL of water was added to dissolve the primary standard, should it be measured by measuring cylinder or pipette? Why?

3. If the primary standard is not dried, will the standardization result of standard solution concentration be higher or lower?

4. If the NaOH solution which is used to standardize the concentration of HCl solution absorbs CO_2, how does it affect the standardization results?

5. The nitrogen content of ammonium salt is not directly determined using NaOH solution. Why?

6. Phenolphthalein is used as the indicator in the neutralization of formic acid in formaldehyde, while methyl is used as the indicator in the neutralization of free acid in ammonium salt solution. Why?

7. Whether formaldehyde method can be used to determine the nitrogen content of NH_4HCO_3?

Risk Assessment and Control Worksheet

TASK	Assessor	Date
Preparation and standardization of NaOH standard solution and Determination of nitrogen content of ammonium salt	Zeying WU, Hongsong WANG, Xingxing WU	1 JUL. 2021

Process Steps	Hazard Identification					Health Risk		Safety Measures				Checklist～Risk Assessment				Action	
	Chemical Name	Class	Concn.	Amount	MSDS	Type of Injury Possible	or Physical Hazard	To reduce or eliminate the risk	Technique	Glasses	Gloves	Fume Cbd.	Risk Score		Risk Level	In case of Emergency/Spill	Special Comments
													Likely 1～4	Hazard 1～4	H M L		

A: Preparation of NaOH saturated solution

Dissolve 120 g of NaOH in 100 mL of water and mix well.	NaOH	8			√	Skin burns and eye injuries		Handle with care		√	√		1	3	L	Sponge & water	

B: Preparation of NaOH solution (0.1 mol·L^{-1})

Transfer 5.6 mL of the supernatant of NaOH saturated solution to a volumetric flask (1 000 mL) and dilute to the mark with water.	NaOH	8	0.1 M	1 000 mL	√	Skin burns and eye injuries		Handle with care		√	√		1	3	L	Sponge & water	

C: Standardization of NaOH solution (0.1 mol·L^{-1})

1) Accurately weigh 0.4～0.5 g of pre-dried KHP into an Erlenmeyer flask (250 mL), add 20～30 mL of water. Heat the flask to dissolve and add 1～2 drops of phenolphthalein indicator. Titrate with 0.1 mol·L^{-1} NaOH standard solution until the color of the solution turns from colorless to reddish and the color stands for 30 s. Record the volume of the NaOH standard solution consumed in the titration.	KHP	5		0.5 g	√	Harmful if swallowed		Handle with care		√	√		1	3	L	Report immediately to Demonstrator or Prep Room	
	NaOH	8	0.1 M	1 000 mL	√	Skin burns and eye injuries		Handle with care		√	√		1	3	L	Sponge & water	
	Phenolphthalein indicator	—	0.5%	1 mol	—								1	1	L	Sponge & water	

Risk Assessment and Control Worksheet

TASK	Assessor	Date
Preparation and standardization of NaOH standard solution and Determination of nitrogen content of ammonium salt	Zeying WU, Hongsong WANG, Xingxing WU	1 JUL. 2021

Process Steps	Hazard Identification				Health Risk		Safety Measures				Risk Assessment				Action	
	Chemical Name	Class or Physical Hazard	Concn.	Amount	MSDS	Type of Injury Possible	To reduce or eliminate the risk	Technique	Glasses	Gloves	Fume Cbd.	Checklist~Risk Assessment	Risk Score		Risk Level	In case of Emergency/Spill	Special Comments
												Likely 1~4	Hazard 1~4		H M L		

II) Repeat 3 times. Calculate the concentration of the NaOH standard solution. Record the experimental data in Table 13 – 1.

D: Preparation of ammonium salt sample

Process Steps	Chemical Name	Class	Concn.	Amount	MSDS	Type of Injury Possible	To reduce or eliminate the risk	Glasses	Gloves	Fume Cbd.	Likely	Hazard	Risk Score	Risk Level	Action
I) Accurately weigh 1.6~1.7 g of ammonium salt, dissolve, transfer to a volumetric flask (250 mL), dilute to the mark with distilled water and shake well. Then transfer 20.00 mL of ammonium salt solution to an Erlenmeyer flask (250 mL).	NH_4Cl	—	0.1 M	250 mL	√	Eye irritant	Handle with care	√	√	√	1	2	2	L	Sponge & water
	$(NH_4)_2SO_4$	—	0.1 M	250 mL	√	—	Handle with care	√	√	√	1	2	2	L	Sponge & water
II) A little of formic acid, contained in formaldehyde solution, can be neutralized by 0.1 mol·L^{-1} NaOH solution using phenolphthalein as the indicator. When the solution is reddish, free acid is removed. Add 5 mL of above formaldehyde solution and 1~2 drops of phenolphthalein to the ammonium salt solution, shake well and stand for 1 min.	HCHO	8	40%	20 mL	√	Skin burns and eye injuries	Handle with care	√	√	√	1	4	4	L	Sponge & water

Risk Assessment and Control Worksheet *continued*

TASK	Preparation and standardization of NaOH standard solution and Determination of nitrogen content of ammonium salt			Assessor	Zeying WU, Hongsong WANG, Xingxing WU				Date	1 JUL. 2021		
Process Steps	Hazard Identification			Health Risk	Safety Measures				Risk Assessment			Action
	Chemical Name	Class or Physical Hazard	Concn.	Amount	MSDS	Type of Injury Possible	To reduce or eliminate the risk	Technique	Glasses	Gloves	Fume Cbd.	Checklist～Risk Assessment	Likely 1～4	Hazard 1～4	Risk Score	Risk Level H M L	In case of Emergency/Spill	Special Comments
E: Titration																	
Experimental Product (s)																	
Experimental Waste (s)																	
Final assessment of overall process	Low risk when correct procedures followed																
Signatures	Supervisor:				Lab Manager/OHS Representative:				Student/Operator:			

实验 14　HCl 标准溶液的配制与标定及混合碱的分析(双指示剂法)

【实验目的】

1. 掌握用无水碳酸钠作为基准物质标定盐酸的方法。
2. 熟悉双指示剂指示的应用和掌握判断滴定终点的方法。
3. 掌握双指示剂法测定混合碱中各组分含量的原理和方法。

【实验原理】

1. HCl 标准溶液的配制与标定

无水碳酸钠(Na_2CO_3)易吸收空气中的水分,使用前需 200~270 ℃干燥 1 h,保存于干燥器中。用其标定盐酸的反应为:

$$Na_2CO_3 + 2HCl = 2NaCl + H_2O + CO_2\uparrow$$

在化学计量点时,溶液为饱和 H_2CO_3 溶液,其 pH 为 3.9,可用甲基橙作指示剂,终点溶液颜色由黄色变为橙色。

2. 混合碱的分析

混合碱是指 Na_2CO_3 与 NaOH 或 $NaHCO_3$ 与 Na_2CO_3 的混合物,测定混合碱中各组分含量时,可选用 HCl 标准溶液进行滴定分析。双指示剂法是根据滴定过程中溶液 pH 的变化,选用两种不同指示剂来分别指示第一、第二计量点。

在混合碱中加入酚酞指示剂,用 HCl 溶液滴定至溶液由红色变为微红色,此时消耗 HCl 溶液的体积为 V_1。此过程发生的反应如下:

$$HCl + NaOH = NaCl + H_2O$$
$$HCl + Na_2CO_3 = NaHCO_3 + H_2O$$

再加入甲基橙指示剂,继续用 HCl 溶液滴定至溶液由黄色变为橙色,此时消耗 HCl 溶液的体积为 V_2。此过程发生的反应如下:

$$HCl + NaHCO_3 = NaCl + H_2O + CO_2\uparrow$$

根据消耗 HCl 溶液的体积 V_1 和 V_2,可分别计算混合碱中 Na_2CO_3 与 NaOH 或 $NaHCO_3$ 与 Na_2CO_3 的含量。

(1) 当 $V_1 > V_2$ 时,试样为 Na_2CO_3 与 NaOH 的混合物。

$$NaOH\% = \frac{C_{HCl} \times (V_1 - V_2) \times 10^{-3} \times M_{NaOH}}{m \times \frac{25}{250}} \times 100$$

$$\mathrm{Na_2CO_3}\% = \frac{C_{\mathrm{HCl}} \times V_2 \times 10^{-3} \times M_{\mathrm{Na_2CO_3}}}{m \times \dfrac{25}{250}} \times 100$$

$$M_{\mathrm{NaOH}} = 400.00 (\mathrm{g/mol}), M_{\mathrm{Na_2CO_3}} = 105.99 (\mathrm{g/mol})$$

(2) 当 $V_1 < V_2$ 时，试样为 $\mathrm{NaHCO_3}$ 与 $\mathrm{Na_2CO_3}$ 的混合物。

$$\mathrm{NaOH}\% = \frac{C_{\mathrm{HCl}} \times (V_2 - V_1) \times 10^{-3} \times M_{\mathrm{NaHCO_3}}}{m \times \dfrac{25}{250}} \times 100$$

$$\mathrm{Na_2CO_3}\% = \frac{C_{\mathrm{HCl}} \times V_1 \times 10^{-3} \times M_{\mathrm{Na_2CO_3}}}{m \times \dfrac{25}{250}} \times 100$$

(3) 当 $V_1 = V_2$ 时，试样为 $\mathrm{Na_2CO_3}$。

【器材与试剂】

1. 器材：25 mL 酸式滴定管，250 mL 容量瓶，500 mL 容量瓶，500 mL 试剂瓶，25 mL 移液管，250 mL 锥形瓶，25 mL 量筒，50 mL 烧杯，电子天平。
2. 试剂：浓 HCl，无水 $\mathrm{Na_2CO_3}$（基准物质），酚酞指示剂，甲基橙指示剂（1 g·L^{-1} 水溶液）。

【实验内容】

1. 0.10 mol·L^{-1} HCl 标准溶液的配制

移取 4.2~4.5 mL 浓盐酸于 50 mL 烧杯中，转移至 500 mL 容量瓶中，用蒸馏水稀释至标线，混合均匀，倒入洁净的试剂瓶中，塞上玻璃塞，密封备用。

2. 0.10 mol·L^{-1} HCl 标准溶液的标定

准确称取 1.0~1.2 g 无水碳酸钠，用蒸馏水溶解，全部转移至 250 mL 容量瓶中，定容，摇匀。用移液管准确移取 25.00 mL 溶液于 250 mL 锥形瓶中，加入 20~30 mL 水溶解，加入 1~2 滴甲基橙指示剂，用 HCl 标准溶液滴定至溶液由黄色变为橙色，即为终点。

3. 混合碱的分析

用移液管准确移取 25.00 mL 混合碱溶液于 250 mL 锥形瓶中，加入 2~3 滴酚酞指示剂，用 0.10 mol·L^{-1} HCl 标准溶液滴定至溶液由红色变为微红色，为第一终点，记录消耗的 HCl 标准溶液体积 V_1，再加入 2 滴甲基橙指示剂，继续用 HCl 标准溶液滴定至溶液由黄色变为橙色，为第二终点，记录第二次消耗的 HCl 标准溶液体积 V_2。平行滴定 3 次。

【注意事项】

1. 基准物质无水碳酸钠容易吸收水和 $\mathrm{CO_2}$，应烘干后置于干燥器中冷却备用。
2. 标定 HCl 溶液时，由于反应本身产生的 $\mathrm{H_2CO_3}$ 会使指示剂颜色变化不够明显，所以在临近滴定终点之前，最好将溶液加热至沸腾，并摇动以除去 $\mathrm{CO_2}$，冷却后再滴至溶液呈橙色为止。
3. 当混合碱为 NaOH 和 $\mathrm{Na_2CO_3}$ 组成时，酚酞指示剂用量可适当多加几滴，否则常因滴定不完全而使 NaOH 的测定结果偏低，$\mathrm{Na_2CO_3}$ 的测定结果偏高。

4. 第一计量点的颜色变化为红色至微红色,不应有 CO_2 的损失,造成 CO_2 损失的操作是滴定速度过快,溶液中 HCl 局部过量。因此滴定速度宜适中,摇动要均匀。

5. 第二计量点时颜色变化为黄色至橙色。滴定过程中摇动要剧烈,使 CO_2 逸出,避免形成碳酸饱和溶液,使终点提前。

【思考题】

1. 为什么 HCl 标准溶液不采用直接法配制?
2. 用双指示剂法测定混合碱组成的方法和原理是什么?
3. 采用双指示剂法测定混合碱,判断下列五种情况下,混合碱的组成。
(1) $V_1=0, V_2>0$　(2) $V_1>0, V_2=0$　(3) $V_1>V_2$　(4) $V_1<V_2$　(5) $V_1=V_2$

● 盐酸浓度标定

风险评估及控制工作表

实验任务	HCl标准溶液的配制与标定及混合碱的分析（双指示剂法）				评估人	吴泽颖、王红松、吴星星			评估日期	2021年7月1日				
过程步骤	危害鉴别				健康风险	降低或消除风险	安全措施			风险评估			措施	
	化合物名称	类别	浓度	用量	化学品安全技术说明书	可能的伤害种类		技术	防护眼镜	手套	通风橱	清单/风险评估	可能性 1~4	危害 1~4	风险	危害级别 高中低	紧急情况/泼洒	特殊备注

A: 0.10 mol·L⁻¹ HCl标准溶液的配制

移取 4.2~4.5 mL 浓盐酸于 50 mL烧杯中，转移至 500 mL 容量瓶中，用蒸馏水稀释至标线，混合均匀，倒入洁净的试剂瓶中，塞上玻璃塞，密封备用。

| | HCl | | 12 M | 4.5 mL | √ | 眼睛、呼吸系统和皮肤刺激 | 关注 | | √ | √ | √ | | 1 | 4 | 4 | 低 | 用碳酸氢钠中和，并及时擦试干净 | |

B: 0.10 mol·L⁻¹ HCl标准溶液的标定

准确称取 1.0~1.2 g 无水碳酸钠，用蒸馏水溶解，全部转移至 250 mL 容量瓶中，定容，摇匀。用移液管准确移取 25.00 mL 溶液于 250 mL 锥形瓶中，加入 20~30 mL水溶解，加入 1~2 滴甲基橙指示剂，用 HCl标准溶液滴定至溶液由黄色变为橙色，即为终点。

	无水碳酸钠			0.2 g	√	眼睛刺激	关注		√	√	√		1	3	3	低	大量水冲洗	
	HCl	2.3+8	0.1 M	500 mL	√	皮肤灼伤和眼损伤	关注		√	√	√		1	3	3	低	大量水冲洗	
	甲基橙指示剂		0.5%	1 mL	—	—			√	√	√		1	1	1	低	大量水冲洗	

C: 混合碱的分析

用移液管准确移取 25.00 mL 混合碱溶液于 250 mL 锥形瓶中，加入 2~3 滴酚酞指示剂，用 0.10 mol·L⁻¹ HCl 标准溶液滴定至红色变为微红色，为第一终点，记录消耗的 HCl 标准溶液体积 V_1，再加入 2 滴甲基橙指示剂，继续用 HCl 标准溶液滴定至黄色变为橙色，为第二终点，记录第二次消耗的 HCl 标准溶液体积 V_2。平行滴定 3 次。

	酚酞指示剂		1%	1 mL	—	—			√	√	√		1	1	1	低	大量水冲洗	
	HCl	2.3+8	0.1 M	500 mL	√	皮肤灼伤和眼损伤	关注		√	√	√		1	3	3	低	大量水冲洗	
	甲基橙指示剂		0.5%	1 mL	—	—			√	√	√		1	1	1	低	大量水冲洗	

实验废弃物

整个过程的最终评估：低风险（按照正确流程操作）

签名   实验老师：   实验室主任/职业安全健康(OHS)代表：   学生/操作人：

Experiment 14 Preparation and standardization of HCl standard solution and determination of the composition of mixed base (double-indicator method)

【Objectives】

1. To master the method of using borax as the primary standard to standardize hydrochloride acid solution.

2. To acquaint with the method of methyl red indicator.

3. To master the principle and method of using double indicator to determine the content of mixed base.

【Principles】

1. Preparation and standardization of HCl standard solution

Anhydrous sodium carbonate (Na_2CO_3) is easy to absorb moisture in the air. Before use, it needs to be dried at 200~270 ℃ for 1 h and stored in a dryer. The reaction of standardization HCl solution with borax is as follows:

$$Na_2CO_3 + 2HCl = 2NaCl + H_2O + CO_2 \uparrow$$

At the stoichiometric point, the solution is saturated H_2CO_3 solution and the pH value of the solution is 3.9, methyl orange can be used as indicator, and the color of the end point changes from yellow to orange.

2. Analysis of mixed base

The mixed base is a mixture of Na_2CO_3 and NaOH or $NaHCO_3$ and Na_2CO_3. HCl standard solution can be used for titration to determine the content of each component in the mixed base. According to the change of pH value of the solution in the titration process, two different indicators are used to indicate the first and second measuring points respectively. This titration method is called "double indicator method". Add phenolphthalein indicator in the mixed base, titrate with HCl standard solution until the color of solution changes from red to reddish. The volume of HCl standard solution consumed is V_1. The reactions are as follows.

$$HCl + NaOH = NaCl + H_2O$$

$$HCl + Na_2CO_3 = NaHCO_3 + H_2O$$

Then add the methyl orange indicator, continue to titrate with HCl standard solution until the color of solution changes from yellow to orange. The volume of HCl standard solution consumed is V_2. The reaction is as follows.

$$HCl + NaHCO_3 = NaCl + H_2O + CO_2 \uparrow$$

The percentage composition of mixed base can be determined according to V_1 and V_2.

(1) If $V_1 > V_2$, the mixed base is a mixture of Na_2CO_3 and NaOH.

$$NaOH\% = \frac{C_{HCl} \times (V_1 - V_2) \times 10^{-3} \times M_{NaOH}}{m \times \frac{25}{250}} \times 100$$

$$Na_2CO_3\% = \frac{C_{HCl} \times V_2 \times 10^{-3} \times M_{Na_2CO_3}}{m \times \frac{25}{250}} \times 100$$

$$M_{NaOH} = 400.00 (g/mol), M_{Na_2CO_3} = 105.99 (g/mol)$$

(2) If $V_1 < V_2$, the mixed base is a mixture of $NaHCO_3$ and Na_2CO_3.

$$NaOH\% = \frac{C_{HCl} \times (V_2 - V_1) \times 10^{-3} \times M_{NaHCO_3}}{m \times \frac{25}{250}} \times 100$$

$$Na_2CO_3\% = \frac{C_{HCl} \times V_1 \times 10^{-3} \times M_{Na_2CO_3}}{m \times \frac{25}{250}} \times 100$$

(3) If $V_1 = V_2$, the mixed base is Na_2CO_3.

【Apparatus and Chemicals】

1. Apparatus

Acid burette (25 mL); Volumetric flask (250 mL); Volumetric flask (500 mL); Reagent bottle (500 mL); Pipette (25 mL); Erlenmeyer flask (250 mL); Measuring cylinder (25 mL); Beaker (50 mL); Electronic balance.

2. Chemicals

Concentrated hydrochloric acid; anhydrous Na_2CO_3 (primary standard); phenolphthalein indicator; methyl orange indicator (1 g·L^{-1} aqueous solution).

【Experimental Procedures】

1. Preparation of HCl solution (0.10 mol·L^{-1})

Transfer 4.2~4.5 mL of concentrated HCl into a beaker (50 mL). Then transfer it to a volumetric flask (500 mL), dilute it to the mark with distilled water and mix well. Store it in a clean reagent bottle with stopper.

2. Standardization of HCl solution (0.10 mol·L^{-1})

Accurately weigh 1.0~1.2 g anhydrous Na_2CO_3 and dissolve it with distilled water. Transfer to a volumetric flask (250 mL), dilute with distilled water to the mark and shake well. Then transfer 25.00 mL of Na_2CO_3 solution to an Erlenmeyer flask (250 mL). Add 20~30 mL of distilled water and 1~2 drops of methyl orange indicator to it, and titrate with HCl standard solution until the color is from yellow to orange.

3. Analysis of mixed base

Transfer 25.00 mL of mixed base solution to an Erlenmeyer flask (250 mL). Add 2~3 drops of phenolphthalein indicator and titrate with HCl standard solution until the color turns from red to reddish which is the first endpoint. Record the volume of HCl standard solution consumed (V_1). Then add 2 drops of methyl orange indicator in the solution and continue to titrate with HCl standard solution until the color turns from yellow to orange which is the second endpoint. Record the volume of HCl standard solution consumed for the second time (V_2). Repeat 3 times.

【Notes】

1. Anhydrous Na_2CO_3 is easy to absorb water and CO_2. It should be dried and placed in a desiccator for cooling.

2. In the standardization of HCl solution, H_2CO_3 generated in the reaction affects the change of color. Consequently, it is better to heat the solution to remove CO_2.

3. When the mixed base is the mixture of NaOH and Na_2CO_3, phenolphthalein indicator should be added more drops. Otherwise the results may be lower because of incomplete titration.

4. The titration arrives at the first endpoint when the color is from red to reddish. There should not be a loss of CO_2. If the titration speed is too fast, local excessive of HCl results in the loss of CO_2. Therefore, the titrate speed must be moderate.

5. During the process between the first endpoint and the second endpoint, shaking must be severe for CO_2 effusion to avoid the formation of saturated carbonate solution and advance endpoint.

【Questions】

1. Why is HCl standard solution not prepared directly?
2. What is the principle of determination the percentage composition of mixed base by double indicator method?
3. Judge the composition of mixed base in the following five cases when double indicator method is used to determine the percentage composition.
 (1) $V_1=0$, $V_2>0$ (2) $V_1>0$, $V_2=0$ (3) $V_1>V_2$ (4) $V_1<V_2$ (5) $V_1=V_2$

Risk Assessment and Control Worksheet

TASK	Preparation and standardization of HCl standard solution and Determination of the composition of mixed base (double-tracer technique)				Assessor	Zeying WU, Hongsong WANG, Xingxing WU				Date	1 JUL. 2021			
Process Steps	Hazard Identification				Health Risk		Safety Measures				Checklist＼Risk Assessment	Risk Assessment		Action
	Chemical Name	Class	Concn.	Amount	MSDS	Type of Injury Possible	To reduce or eliminate the risk	Technique	Glasses	Gloves	Fume Cbd.	Likely	Hazard	Risk Score	Risk Level	In case of Emergency/Spill
					or Physical Hazard							$1\sim4$	$1\sim4$		H M L	Special Comments

A: Preparation of HCl solution (0.10 mol·L^{-1})

Transfer 4.2~4.5 mL of concentrated HCl into a beaker (50 mL). Then transfer it to a volumetric flask (500 mL), dilute it to the mark with distilled water and mix well. Store it in a clean reagent bottle with stopper.

| HCl | | 12 M | 4.5 mL | | Eye, respiratory system & skin irritant | Handle with care | | √ | √ | √ | 1 | 4 | 4 | L | Neutralise with sodium bicarbonate, sponge & water. |

B: Standardization of HCl solution (0.10 mol·L^{-1})

Accurately weigh 1.0 ~ 1.2 g anhydrous Na$_2$CO$_3$ and dissolve it with distilled water. Transfer to a volumetric flask (250 mL), dilute with distilled water to the mark and shake well. Then transfer 25.00 mL of Na$_2$CO$_3$ solution to an Erlenmeyer flask (250 mL). Add 20~30 mL of distilled water and 1~2 drops of methyl orange indicator to it, and titrate with HCl standard solution until the color is from yellow to orange.

Anhydrous Na$_2$CO$_3$	2.3+8		0.2 g	√	Eye irritant	Handle with care		√	√		1	3	3	L	Sponge & water
HCl		0.1 M	500 mL	√	Skin burns and eye injuries	Handle with care		√	√		1	3	3	L	Sponge & water
Methyl orange indicator	—	0.5%	1 mL	—	—	Handle with care		√	√		1	1	1	L	Sponge & water

continued

Risk Assessment and Control Worksheet

TASK	Assessor	Date
Preparation and standardization of HCl standard solution and Determination of the composition of mixed base (double-tracer technique)	Zeying WU, Hongsong WANG, Xingxing WU	1 JUL. 2021

Process Steps	Hazard Identification				Health Risk	Safety Measures				Checklist↘Risk Assessment			Action	
	Chemical Name	Class or Physical Hazard	Concn.	Amount	MSDS	Type of Injury Possible	To reduce or eliminate the risk	Technique	Glasses	Gloves	Fume Cbd.	Likely 1~4	Hazard 1~4	Risk Score	Risk Level H M L	In case of Emergency/Spill	Special Comments

C: Analysis of mixed base

Process Steps	Chemical	Class	Concn.	Amount	MSDS	Injury	Technique	Glasses	Gloves	Fume	Likely	Hazard	Score	Level	Emergency	Comments
Transfer 25.00 mL of mixed base solution to an Erlenmeyer flask (250 mL). Add 2~3 drops of phenolphthalein indicator and titrate with HCl standard solution until the color turns from red to reddish which is the first endpoint. Record the volume of HCl standard solution consumed (V_1). Then add 2 drops of methyl orange indicator in the solution and continue to titrate with HCl standard solution until the color turns from yellow to orange which is the second endpoint. Record the volume of HCl standard solution consumed for the second time (V_2). Repeat 3 times.	Phenolphthalein indicator	—	1%	1 mL	—	—	Handle with care	√	√		1	3	3	L	Sponge & water	
	HCl	2.3+8	0.1 M	500 mL	√	Skin burns and eye injuries	Handle with care	√	√		1	3	3	L	Sponge & water	
	Methyl orange indicator	—	0.5%	1 mL	—	—	Handle with care	√	√		1	1	1	L	Sponge & water	

Experimental Product(s)	
Experimental Waste(s)	
Final assessment of overall process	Low risk when correct procedures followed
Signatures	Supervisor: Lab Manager/OHS Representative: Student/Operator:

实验 15　EDTA 标准溶液的配制与标定及水硬度的测定

【实验目的】

1. 掌握 EDTA 标准溶液的配制和标定方法。
2. 掌握铬黑 T 指示剂指示终点的方法。
3. 掌握配位滴定法测定水中钙、镁的方法。
4. 了解水的硬度的计算。

【实验原理】

1. EDTA 标准溶液的配制与标定

乙二胺四乙酸(简称 EDTA)，分子中有六个可能的位置可以与金属成键：四个羧基氧和两个氨基氮。EDTA 的存在形式包括：H_4Y，H_3Y^-，H_2Y^{2-}，HY^{3-} 和 Y^{4-}。HY_4 难溶于水，故 EDTA 标准溶液通常用乙二胺四乙酸二钠盐($Na_2H_2Y \cdot 2H_2O$)来配制。乙二胺四乙酸二钠盐很难得到纯物质，其标准溶液常用间接法配制。

EDTA 能与大多数金属离子形成 1∶1 的稳定配合物，通常可用含量不低于 99.95% 的某些金属（如：Cu、Zn、Ni、Pb 等）及它们的金属氧化物或某些盐类（如 $ZnSO_4 \cdot 7H_2O$、$MgSO_4 \cdot 7H_2O$、$CaCO_3$ 等）作为基准物质来标定 EDTA 溶液。本实验选用 $CaCO_3$ 为基准物质以标定 EDTA，选用钙指示剂指示终点，其变色原理为：

滴定前：$Ca^{2+} + HIn^{2-}$（蓝）$= CaIn^-$（酒红色）$+ H^+$

滴定中：$Ca^{2+} + H_2Y^{2-} = CaY^{2-} + 2H^+$

终点时：$CaIn^-$（酒红色）$+ H_2Y^{2-} = CaY^{2-} + HIn^{2-}$（蓝色）$+ H^+$

2. 水硬度的测定

水的硬度是表示水质的一项重要指标。水的总硬度是指水中 Ca^{2+}、Mg^{2+} 的总量。

水的总硬度测定主要采用 EDTA 滴定法。在 pH≈10 的氨性缓冲溶液中，以铬黑 T(EBT)为指示剂，用 EDTA 标准溶液直接滴定 Ca^{2+}、Mg^{2+} 的总量。EDTA 和金属指示剂铬黑 T(H_3In)分别与 Ca^{2+}、Mg^{2+} 形成络合物，稳定性为 $CaY^{2-} > MgY^{2-} > MgIn^- > CaIn^-$，当水样中加入少量铬黑 T 指示剂时，它首先和 Mg^{2+} 生成红色络合物 $MgIn^-$，然后与 Ca^{2+} 生成红色络合物 $CaIn^-$。当 EDTA 滴入时，EDTA 与 Ca^{2+}、Mg^{2+} 络合，滴定终点时 EDTA 竞争络合物 $CaIn^-$、$MgIn^-$ 中的 Ca^{2+}、Mg^{2+}，将 EBT 置换出来，溶液颜色由酒红色变为蓝色。

$$CaIn^- + H_2Y^{2-} = CaY^{2-} + HIn^{2-} + H^+$$

$$MgIn^- + H_2Y^{2-} = MgY^{2-} + HIn^{2-} + H^+$$

　　酒红色　　　　　　　　　　蓝色

滴定时 Fe^{3+}、Al^{3+} 的干扰可用三乙醇胺掩蔽，Cu^{2+}、Pb^{2+}、Zn^{2+} 等重金属离子可用 KCN、Na_2S 将其掩蔽。如需分别测定水的钙硬度和镁硬度，可通过加入氢氧化钠，在 pH≈12 的溶液中测定钙离子，这样可以通过强碱使钙离子和氢氧化镁沉淀分离。继而用钙试剂作为指示剂指示终点(紫红色～蓝色)。

【器材与试剂】

1. 器材：50 mL 酸式滴定管，250 mL 容量瓶，500 mL 容量瓶，500 mL 试剂瓶，25 mL 移液管，250 mL 锥形瓶，100 mL 量筒，100 mL 烧杯，500 mL 烧杯，表面皿，电子天平。

2. 试剂：乙二胺四乙酸二钠盐，$CaCO_3$ 优级纯(110 ℃ 干燥至恒重，干燥器中保存)，0.5% 铬黑 T 指示剂(0.5 g 铬黑 T，加 20 mL 三乙醇胺，用水稀释至 100 mL)，1% 钙指示剂(1 g 钙指示剂和 100 g NaCl 研细混匀，储存于干燥器中)，氨性缓冲溶液(pH≈10，20 g NH_4Cl 和 100 mL 浓氨水，用水稀释至 1 L)，6 mol·L^{-1} HCl 溶液，1 mol·L^{-1} NaOH 溶液。

【实验内容】

1. 0.02 mol·L^{-1} EDTA 标准溶液的配制

称取 3.0 g 乙二胺四乙酸二钠盐于 500 mL 烧杯中加 200 mL 水，溶解完全后转入聚乙烯试剂瓶中，用水稀释至 400 mL，摇匀。

2. 0.02 mol·L^{-1} 钙标准溶液的配制

准确称取干燥过的 $CaCO_3$ 0.4～0.5 g，用少量水润湿，盖上表面皿，慢慢滴加 6 mol·L^{-1} HCl 溶液 5 mL 使其溶解，定量转移至 250 mL 容量瓶中，定容、摇匀，计算其准确浓度。

$$C_{Ca^{2+}} = \frac{m \times 1\,000}{M \times 250} \text{ mol·L}^{-1} \quad (M=100.09)$$

3. EDTA 溶液浓度的标定

准确移取 25.00 mL 钙标准溶液于锥形瓶中，加 5 mL 1 mol·L^{-1} NaOH 溶液和适量钙指示剂，摇匀，用待标定的 EDTA 溶液进行滴定，直至溶液由紫红色恰好变为纯蓝色，即为终点，记录所消耗的 EDTA 溶液的体积 V_Y。平行滴定 3 次。

4. 水样总硬度的测定

用移液管移取适量水样于锥形瓶中(自来水取样 100.00 mL；人工配制水样取 25.00 mL)，加 5 mL 氨性缓冲溶液和适量铬黑 T 指示剂(3～4 滴)，摇匀，用 EDTA 标准溶液进行滴定。因反应较慢，临近终点时应缓慢滴加，充分振摇，直至溶液颜色由紫红色恰好变为纯蓝色，即为终点。记录 EDTA 的用量 V_1。平行滴定 3 次。

5. 钙硬度的测定

移取等体积水样于锥形瓶中，加入适量 1 mol·L^{-1} NaOH 溶液(自来水样中加 2 mL；人工配制水样中加 5 mL)，使水样 pH 调整为 12～13，加入适量钙指示剂，用 EDTA 标准溶液进行滴定，并不断振荡。临近终点时应缓慢滴加，充分振摇，直至溶液颜色由紫红色恰好变为纯蓝色，即为终点。记录 EDTA 的用量 V_2。平行滴定 3 次。

6. 镁硬度的计算

由水的总硬度及水的钙硬度可算出水的镁硬度。

【注意事项】

1. EDTA 溶液应保存在聚乙烯或硬质玻璃瓶中,以免 EDTA 与玻璃中的金属离子反应。

2. 铬黑 T 与 Mg^{2+} 显色的灵敏度高,与 Ca^{2+} 显色的灵敏度低,当水样中钙含量很高而镁含量很低时,指示终点的颜色变化不很敏锐。可在水样中加入少许 MgY^{2-} 溶液以提高终点变色的敏锐性,或采用 K-B 混合指示剂指示终点(紫红色至蓝绿色)。

【思考题】

1. 络合滴定中为什么加入缓冲溶液?
2. 用 Na_2CO_3 为基准物,以钙指示剂为指示剂标定 EDTA 浓度时,应控制溶液的酸度为多大?为什么?如何控制?
3. 络合滴定法与酸碱滴定法相比,有哪些不同点?操作中应注意哪些问题?
4. 什么叫水的总硬度?怎样计算水的总硬度?
5. 为什么滴定 Ca^{2+}、Mg^{2+} 总量时要控制 pH≈10,而滴定 Ca^{2+} 分量时要控制 pH 为 12~13?pH>13 时,对测 Ca^{2+} 含量有何影响?
6. 如果只有铬黑 T 指示剂,能否测定 Ca^{2+} 的含量?如何测定?

● 视频演示

风险评估及控制工作表

EDTA 标准溶液的配制与标定及水硬度的测定

评估人	吴泽颖、王红松、吴星星	评估日期	2021 年 7 月 1 日

实验任务：EDTA 标准溶液的配制与标定及水硬度的测定

过程步骤	危害鉴别			健康风险		安全措施			评估清单/风险评估	风险评估			措施	
化合物名称	类别	浓度	用量	化学品安全技术说明书	可能的伤害种类	降低或消除风险	技术	防护眼镜	手套	通风橱	可能性 1~4	危害 1~4	风险	危害级别 高/中/低	紧急情况/泼洒 特殊备注

A：0.02 mol·L⁻¹ EDTA 标准溶液的配制

过程步骤：称取 3.0 g 乙二胺四乙酸二钠盐于 500 mL 烧杯中，加 200 mL 水，温热使其溶解完全，转入聚乙烯试剂瓶中，用水稀释至 400 mL，摇匀。

| 乙二胺四乙酸二钠盐 | — | 0.02 M | 400 mL | √ | 皮肤/眼刺激 | 关注 | | √ | √ | | 1 | 1 | 1 | 低 | 大量水冲洗 |

B：0.02 mol·L⁻¹ 钙标准溶液的配制

过程步骤：准确称取干燥过的 CaCO₃ 0.4~0.5 g，用少量水润湿，盖上表面皿，慢慢滴加 HCl（1:1）5 mL 使其溶解，定量转移至 250 mL 容量瓶中，定容，摇匀，计算其准确浓度。

| HCl | 2.3+8 | 6 M | 5 mL | √ | 皮肤灼伤和眼损伤 | 关注 | | √ | √ | √ | 1 | 3 | 3 | 低 | 大量水冲洗 |

C：EDTA 溶液浓度的标定

D：水样总硬度的测定

过程步骤：用移液管移取适量水样于锥形瓶中（自来水取 100.00 mL），人工配制水取 25.00 mL，加入 5 mL 氨性缓冲溶液和适量铬黑 T 指示剂（3~4 滴），摇匀，用 EDTA 标准溶液进行滴定。因反应较慢，临近终点时应缓慢滴加，充分振摇，直至溶液颜色由紫红色恰好变为纯蓝色，即为终点。记录 EDTA 的用量 V_1。平行滴定 3 次。

| NH₃·H₂O | 8 | 2 M | 5 mL | √ | 皮肤灼伤和眼损伤 | 关注 | | √ | √ | √ | 1 | 4 | 4 | 低 | 大量水冲洗 |
| 铬黑T指示剂 | — | — | 1 mL | — | 眼睛刺激 | 关注 | | √ | √ | | 1 | 1 | 1 | 低 | 大量水冲洗 |

续 表

风险评估及控制工作表

实验任务	EDTA 标准溶液的配制与标定及水硬度的测定				评估人	吴泽颖、王红松、吴星星			评估日期	2021 年 7 月 1 日		
过程步骤	危害鉴别				健康风险		安全措施		清单/风险评估	风险评估		紧急情况/泼洒措施
	化合物名称	类别	浓度	用量	化学品安全技术说明书	可能的伤害种类	降低或消除风险	技术 防护眼镜 手套 通风橱		可能性 危害	风险 危害级别	
		或物理危害								1~4 1~4	高中低	特殊备注
E: 钙硬度的测定												
移取等体积水样于锥形瓶中,加入适量 1 mol·L⁻¹ NaOH 溶液(自来水样中加 2 mL;人工配制水样中加 5 mL),使水样 pH 调整为 12～13,加入适量钙指示剂,用 EDTA 标准溶液进行滴定,并不断振荡。临近终点时应缓慢滴加,充分振摇,直至溶液颜色由紫红色恰好变为纯蓝色,即为终点。记录 EDTA 的用量 V_2。平行滴定 3 次。	NaOH	8	1 M	5 mL	√	皮肤灼伤和眼损伤	关注	√ √		1 3	3 低	大量水冲洗
F: 由水的总硬度及水的钙硬度可算出水的镁硬度												
实验产物												
实验废弃物												
整个过程的最终评估	低风险(按照正确流程操作)											
签名	实验老师:						实验室主任/职业安全健康(OHS)代表:			学生/操作人:		

Experiment 15 Preparation and standardization of EDTA standard solution and determination of water hardness

【Objectives】

1. To master the method of preparation and standardization of standard EDTA solution.
2. To master the way to determine the end point of Eriochrome Black T (EBT) indicator.
3. To master the EDTA titration method of Ca^{2+} and Mg^{2+} in natural water.
4. Understand the calculation method of the water hardness.

【Principles】

1. Preparation and standardization of EDTA standard solution

The molecule of ethylenediaminetetraacedic acid which commonly shortened to EDTA has six potential sites for bonding with a metal ion: the four carboxyl groups and the two amino groups. The various EDTA species are often abbreviated H_4Y, H_3Y^-, H_2Y^{2-}, HY^{3-} and Y^{4-}. H_4Y dissolve in water slightly. So, EDTA standard solution is commonly prepared by dissolving disodium dihydrogen ethylenediamineteraacetate ($Na_2H_2Y \cdot 2H_2O$) in water. It is not easy to get the pure product. The standard solution needs to be prepared by indirect method.

EDTA can form 1:1 stable complexes with most metal ions. In general, some metals (such as Cu, Zn, Ni, Pb, etc.) with content of no less than 99.95% and their metal oxides or some salts (such as $ZnSO_4 \cdot 7H_2O$, $MgSO_4 \cdot 7H_2O$, $CaCO_3$, etc.) can be used as primary standard to standardize EDTA solution. In this experiment, $CaCO_3$ is selected as the primary standard to standardize EDTA, and calcium indicator is used to indicate the end point. The color changing principle is as follows:

Before titration: $Ca^{2+} + HIn^{2-}$ (blue) $== CaIn^-$ (wine red) $+ H^+$

In titration: $Ca^{2+} + H_2Y^{2-} == CaY^{2-} + 2H^+$

End point: $CaIn^-$ (wine red) $+ H_2Y^{2-} == CaY^{2-} + HIn^{2-}$ (blue) $+ H^+$

2. Determination of water hardness

The hardness of water is an important indicator of water quality. The total hardness of water refers to the total amount of Ca^{2+} and Mg^{2+} in water.

The total hardness of water is mainly determined by EDTA titration. The concentration of Ca^{2+} and Mg^{2+} is determined by titration using EDTA standard solution and Eriochrome Black T (EBT) as the indicator In ammonia buffer solution with pH≈10. EDTA and metal indicator EBT (H_3In) form complexes with Ca^{2+} and Mg^{2+} respectively, with the stability of $CaY^{2-} > MgY^{2-} > MgIn^- > CaIn^-$. The small amount of EBT added to the water sample first form red complex $MgIn^-$ with Mg^{2+}, and then form red complex $CaIn^-$ with Ca^{2+}. When EDTA drips in, EDTA form complexes with Ca^{2+} and Mg^{2+}. At the end point, EDTA competed for complexation of Ca^{2+} and Mg^{2+} in $CaIn^-$ and $MgIn^-$ to replace EBT, and the color of solution changed from wine red to blue.

$$CaIn^- + H_2Y^{2-} \Longrightarrow CaY^{2-} + HIn^{2-} + H^+$$

$$MgIn^- + H_2Y^{2-} \Longrightarrow MgY^{2-} + HIn^{2-} + H^+$$
$$\text{wine red} \qquad\qquad\qquad \text{blue}$$

The interference of Fe^{3+} and Al^{3+} in titration can be masked by triethanolamine, and heavy metal ions such as Cu^{2+}, Pb^{2+} and Zn^{2+} can be masked by KCN and Na_2S.

If the calcium hardness and magnesium hardness of water need to determine respectively, the calcium ion can be determined in the solution with pH $\approx$ 12 by adding sodium hydroxide. In this way, the calcium ion and magnesium hydroxide can be precipitated and separated by strong base. Then calcium reagent was used as indicator to indicate the end point (purplish red to blue).

【Apparatus and Chemicals】

1. Apparatus

Acid burette (50 mL); Volumetric flask (250 mL); Volumetric flask (500 mL); Reagent bottle (500 mL); Pipette (25 mL); Erlenmeyer flask (250 mL); Measuring cylinder (100 mL); Beaker (100 mL); Beaker (500 mL); Watch glass; Electronic balance.

2. Chemicals

Disodium ethylenediaminetetraacetate; $CaCO_3$ (GR, dried to constant weight at 110 ℃; stored in a desiccator); 0.5% EBT indicator (0.5 g EBT, added 20 mL triethanolamine, diluted to 100 mL with water); 1% calcium indicator (1 g calcium indicator and 100 g NaCl are well mixed and stored in a desiccator); ammonia buffer solution (pH $\approx$ 10, 20 g NH_4Cl and 100 mL concentrated ammonia water, dilute to 1 L with water); 6 mol·L^{-1} HCl solution; 1 mol·L^{-1} NaOH solution.

【Experimental Procedures】

1. Preparation of 0.02 mol·L^{-1} EDTA standard solution

Weigh 3.0 g of $Na_2H_2Y \cdot 2H_2O$ in a beaker (500 mL), dissolve in 200 mL of water, dilute to 400 mL with water in warm state and mix well.

2. Preparation of 0.02 mol·L^{-1} calcium standard solution

Accurately weigh 0.4~0.5 g of dried $CaCO_3$, moisten it with a small amount of water, cover the watch glass, slowly add 5 mL of 6 mol·L^{-1} HCl solution to dissolve it, quantitatively transfer it to a 250 mL volumetric flask, volume it and shake it well, and calculate its accurate concentration.

$$C_{Ca^{2+}} = \frac{m \times 1\,000}{M \times 250} \text{ mol·L}^{-1} \quad (M = 100.09)$$

3. Standardization of EDTA standard solution

Accurately transfer 25.00 mL of calcium standard solution into an Erlenmeyer flask, add 5 mL of

1 mol·L^{-1} NaOH solution and a proper amount of calcium indicator. Titrate with EDTA solution until the color of the solution changes from purplish red to pure blue. Record the volume of the EDTA standard solution consumed in the titration (V_Y). Repeat 3 times.

4. Determination of the water hardness

Use a pipette to take a proper amount of water sample and put it into an Erlenmeyer flask (tap water sampling 100.00 mL; manually prepared water sampling 25.00 mL), add 5 mL of ammonia buffer solution and an appropriate amount of EBT indicator (3~4 drops), shake well, and titrate with EDTA standard solution. Due to the slow reaction speed, the solution should be added slowly near the end point and shake well until the solution color changes from purplish red to blue, which is the end point. Record the volume of the EDTA standard solution consumed in the titration (V_1). Repeat 3 times.

5. Determination of the calcium hardness

Transfer the same volume of water sample into an Erlenmeyer flask, add an appropriate amount of 1 mol·L^{-1} NaOH solution (2 mL in tap water sample; 5 mL in artificially prepared water sample), and adjust the pH of the water sample to 12~13. Add an appropriate amount of calcium indicator. Titrate with EDTA standard solution and shake continuously. When it is near the end point, it should be added slowly and shake well until the color of the solution changes from purplish red to blue, which is the end point. Record the volume of the EDTA standard solution consumed in the titration (V_2). Repeat 3 times.

6. Calculation of the magnesium hardness

The magnesium hardness of water can be calculated from the total hardness and the calcium hardness of water.

【Notes】

1. The EDTA solution should be stored in a polyethylene or rigid glass bottle to prevent EDTA from reacting with metal ions in the glass.

2. The color sensitivity of EBT with Mg^{2+} is high, while the sensitivity with Ca^{2+} is low. If the calcium content in the water sample is very high while the magnesium content is very low, the color change indicating the endpoint is not very sharp. Add a little MgY^{2-} solution into the water sample can improve the sensitivity of the endpoint's discoloration, or use a K-B mixed indicator can help in indicating the endpoint (purplish red to blue-green).

【Questions】

1. Why to add buffer solution in complexometric titration?

2. How much acidity should be controlled if Na_2CO_3 was used as primary standard and calcium indicator was used as indicator to titrate EDTA solution? Why? How to control?

3. What are the differences between complexometric titration and acid-base titration? What should be paid attention to in operation?

4. What is the total hardness of water? How to calculate the total hardness of water?

5. Why should we use a pH≈10 buffer for total water hardness titration while using a pH=12~13 buffer for the calcium? What is the effect of calcium hardness if using a pH>13 buffer?

6. Can the content of Ca^{2+} be determined by EBT indicator? How to measure?

Risk Assessment and Control Worksheet

TASK	Preparation and standardization of EDTA standard solution and Determination of the water hardness	Assessor	Zeying WU, Hongsong WANG, Xingxing WU	Date	1 JUL. 2021

Process Steps	Hazard Identification				Health Risk		Safety Measures				Checklist～Risk Assessment			Risk Assessment			Action	
	Chemical Name	Class	Concn.	Amount	MSDS	Type of Injury Possible	To reduce or eliminate the risk	Technique	Glasses	Gloves	Fume Cbd.	Likely	Hazard	Risk Score	Risk Level H M L	In case of Emergency/Spill	Special Comments
												1~4	1~4				

A: Preparation of 0.02 mol·L^{-1} EDTA standard solution

| Weigh 3.0 g of $Na_2H_2Y\cdot2H_2O$ in a beaker (500 mL), dissolve in 200 mL of water, dilute to 400 mL with water in warm state and mix well. | Ethylenediam-inetetraacetic acid disodium salt | — | 0.02 M | 400 mL | √ | Skin and eye irritant | Handle with care | | √ | √ | √ | 1 | 1 | 1 | L | Sponge & water | |

B: Preparation of 0.02 mol·L^{-1} calcium standard solution

| Accurately weigh 0.4~0.5 g of dried $CaCO_3$, moisten it with a small amount of water, cover the watch glass, slowly add 5 mL of HCl (1:1) to dissolve it, quantitatively transfer it to a 250 mL volumetric flask, volume it and shake it well, and calculate its accurate concentration. | HCl | 2,3+8 | 6 M | 5 mL | √ | Skin burns and eye injuries | Handle with care | | √ | √ | √ | 1 | 3 | 3 | L | Sponge & water | |

C: Standardization of EDTA standard solution

continued

Risk Assessment and Control Worksheet

TASK	Preparation and standardization of EDTA standard solution and Determination of the water hardness				Assessor	Zeying WU, Hongsong WANG, Xingxing WU				Date	1 JUL. 2021			
Process Steps	Hazard Identification				Health Risk		Safety Measures				Checklist╲Risk Assessment	Risk Assessment		Action
	Chemical Name	Class	Concn.	Amount	MSDS	Type of Injury Possible	To reduce or eliminate the risk	Technique	Glasses	Gloves	Fume Cbd.	Likely	Hazard	Risk Score	Risk Level H M L	In case of Emergency/Spill
	or Physical Hazard											$1\sim4$	$1\sim4$			Special Comments
	Ammonia	8	2 M	5 mL	√	Skin burns and eye injuries	Handle with care		√	√	√	1	4	4	L	Sponge & water
	EBT indicator	—	—	1 mL	—	Eye irritant	Handle with care		√	√		1	1	1	L	Sponge & water

D: Determination of the water hardness

Use a pipette to take a proper amount of water sample and put it into an Erlenmeyer flask (tap water sampling 100.00 mL; manually prepared water sampling 25.00 mL), add 5 mL of ammonia buffer solution and an appropriate amount of EBT indicator ($3\sim4$ drops), shake well, and titrate with EDTA standard solution. Due to the slow reaction speed, the solution should be added slowly near the end point and shake well until the solution color changes from purplish red to blue, which is the end point. Record the volume of the EDTA standard solution consumed in the titration (V_1). Repeat 3 times.

Risk Assessment and Control Worksheet _continued_

TASK	Preparation and standardization of EDTA standard solution and Determination of the water hardness					Assessor	Zeying WU, Hongsong WANG, Xingxing WU				Date	1 JUL. 2021			
Process Steps	Hazard Identification					Health Risk	Safety Measures				Checklist～Risk Assessment	Risk Assessment			Action
	Chemical Name	Class or Physical Hazard	Concn.	Amount	MSDS	Type of Injury Possible	To reduce or eliminate the risk	Technique	Glasses	Gloves	Fume Cbd.		Likely	Hazard	Risk Score	Risk Level H M L	In case of Emergency/Spill Special Comments
												1～4	1～4			
	Sodium hydroxide	8	1 M	5 mL	✓	Skin burns and eye injuries	Handle with care		✓	✓		1	3	3	L	Sponge & water

E: Determination of the calcium hardness

Transfer the same volume of water sample into an Erlenmeyer flask, add an appropriate amount of 1 mol·L^{-1} NaOH solution (2 mL in tap water sample; 5 mL in artificially prepared water sample), and adjust the pH of the water sample to 12～13. Add an appropriate amount of calcium indicator. Titrate with EDTA standard solution and shake continuously. When it is near the end point, it should be added slowly and shake well until the color of the solution changes from purplish red to blue, which is the end point. Record the volume of the EDTA standard solution consumed in the titration (V_2). Repeat 3 times.

continued

Risk Assessment and Control Worksheet

TASK	Preparation and standardization of EDTA standard solution and Determination of the water hardness				Assessor	Zeying WU, Hongsong WANG, Xingxing WU			Date	1 JUL. 2021			
Process Steps	Hazard Identification				Health Risk	Safety Measures				Checklist～Risk Assessment	Risk Assessment		Action
	Chemical Name	Class or Physical Hazard	Concn.	Amount	MSDS	Type of Injury Possible	To reduce or eliminate the risk	Technique	Glasses	Gloves	Fume Cbd.	Likely	Hazard	Risk Score	Risk Level	In case of Emergency/Spill
												1～4	1～4		H M L	Special Comments

F: **The magnesium hardness of water can be calculated from the total hardness and the calcium hardness of water.**

Experimental Product (s)	
Experimental Waste (s)	
Final assessment of overall process	Low risk when correct procedures followed
Signatures	Supervisor: Lab Manager/OHS Representative: Student/Operator:

实验 16 EDTA 标准溶液的配制与标定及铅、铋混合液中铅、铋含量的连续测定

【实验目的】

1. 掌握 EDTA 标准溶液的配制和标定方法。
2. 掌握二甲酚橙指示剂指示终点的方法。
3. 理解用控制溶液酸度的方法提高 EDTA 选择性的原理。
4. 掌握用 EDTA 进行连续滴定多种金属离子混合溶液的方法。

【实验原理】

1. EDTA 标准溶液的配制与标定

EDTA 标准溶液仍以间接法进行配制。

在选择基准物质对 EDTA 标准溶液进行标定时,除可用实验 15 用到的 $CaCO_3$,还可用 $ZnSO_4 \cdot 7H_2O$。

用 $ZnSO_4 \cdot 7H_2O$ 为基准物质标定 EDTA 时,选用二甲酚橙(XO)作指示剂,以盐酸-六亚甲基四胺调节溶液 pH 为 5~6。其终点反应式为:

$$Zn-XO(紫红色) + Y \Longrightarrow ZnY + XO(黄色)$$

2. 铅、铋混合液中铅、铋含量的连续测定

Bi^{3+}、Pb^{2+} 均能与 EDTA 形成稳定的 1:1 络合物,其 $\lg K$ 值分别为 27.94 和 18.04,由于两者的 $\lg K$ 值相差很大,$\Delta \lg K > 6$,所以可以利用酸效应,通过控制不同的酸度,对其分别进行滴定。在滴定中,以二甲酚橙为指示剂,先用 HNO_3 调节溶液的酸度至 $pH \approx 1.0$,用 EDTA 滴定 Bi^{3+},然后用六亚甲基四胺调节溶液的酸度至 pH 为 5~6,再用 EDTA 继续滴定 Pb^{2+}。连续滴定终点时溶液的颜色变化分别如下。

$pH = 1.0$ 时:$Bi^{3+} + XO \Longrightarrow Bi-XO$
　　　　　　　黄色　　紫红色
　　　　$Bi^{3+} + Y \Longrightarrow BiY$
　　　　$Bi-XO + Y \Longrightarrow BiY + XO$
　　　　紫红色　　　　　　黄色

$pH = 5~6$ 时:$Pb^{2+} + XO \Longrightarrow Pb-XO$
　　　　　　　　黄色　　紫红色
　　　　$Pb^{2+} + Y \Longrightarrow PbY$
　　　　$Pb-XO + Y \Longrightarrow PbY + XO$
　　　　紫红色　　　　　　黄色

【器材与试剂】

1. 器材:50 mL 碱式滴定管,100 mL 容量瓶,250 mL 容量瓶,500 mL 容量瓶,10 mL 移液管,250 mL 锥形瓶,100 mL 烧杯,500 mL 烧杯,电子天平。
2. 试剂:乙二胺四乙酸二钠盐,$ZnSO_4 \cdot 7H_2O$ 基准试剂,0.2% 二甲酚橙溶液,20% 六亚甲基四胺溶液,0.1 mol·L^{-1} HNO_3 溶液,1∶1 HCl 溶液,1∶5 HCl 溶液,Bi^{3+}-Pb^{2+} 混合溶液(其中 Bi^{3+}、Pb^{2+} 浓度各为 0.01 mol·L^{-1},HNO_3 浓度约为 0.15 mol·L^{-1})。

【实验内容】

1. 0.02 mol·L^{-1} EDTA 标准溶液的配制

同实验 15。

2. 0.02 mol·L^{-1} 锌标准溶液的配制

准确称取 $ZnSO_4 \cdot 7H_2O$ 基准试剂 1.40~1.45 g 于 100 mL 小烧杯中,加入约一半的水溶解后,定量转入 250 mL 容量瓶中,稀释,定容,摇匀。

3. EDTA 标准溶液浓度的标定

准确移取 25.00 mL 锌标准溶液于锥形瓶中,加入 2.5 mL 1∶5 HCl 溶液,二甲酚橙指示剂 2 滴,滴加六亚甲基四胺溶液至溶液呈紫红色后,再过量 5 mL,摇匀。用待标定的 EDTA 溶液进行滴定,至溶液由紫红色恰好变为亮黄色为终点,记录 V_Y。平行标定 3 次。

4. Bi^{3+}-Pb^{2+} 的连续测定

用移液管移取 25.00 mL Bi^{3+}-Pb^{2+} 混合液于 250 mL 锥形瓶中,加入 10 mL 0.1 mol·L^{-1} HNO_3 溶液,滴加 2 滴二甲酚橙,用 EDTA 标准溶液滴定至溶液由紫红色突变为亮黄色,即为终点,记录消耗 EDTA 标准溶液的用量 V_1,继续滴加六亚甲基四胺溶液至溶液呈紫红色后,再过量 5 mL,此时溶液的 pH 应为 5~6,然后用 EDTA 标准溶液滴定至溶液再次由紫红色突变为亮黄色,记录消耗 EDTA 标准溶液的用量 V_2。平行滴定 3 次。

【注意事项】

1. 由于 Bi^{3+} 与 EDTA 反应速度较慢,临近终点时应缓慢滴加,充分振摇。
2. Bi^{3+} 极易水解,配制的混合试液中必须具有较高浓度的 HNO_3,临用前再用水稀释至 0.15 mol·L^{-1} 左右。滴定前和滴定初期,不要多用水冲洗锥形瓶口,以防 Bi^{3+} 水解。
3. 标定 EDTA 标准溶液和测定 Pb^{2+} 含量时,终点不明显可加热至 50~60 ℃,使终点易于辨别。测 Pb^{2+} 含量时,在滴定 Bi^{3+} 后调节 pH,滴加 EDTA 标准溶液至近终点再加热,然后小心滴至终点。否则,可能出现白色沉淀,影响终点的判断。

【思考题】

1. 络合滴定中,准确分别滴定的条件是什么?
2. 按本实验操作,滴定 Bi^{3+} 的起始酸度是否超过滴定 Bi^{3+} 的最高酸度?滴定至 Bi^{3+} 的终点时,溶液中酸度为多少?此时再加入 10 mL 200 g·L^{-1} 六亚甲基四胺后,溶液 pH 约为多少?

3. 能否取等量混合试液两份，一份控制 pH≈1.0 滴定 Bi^{3+}，另一份控制 pH 为 5～6 滴定 Bi^{3+}、Pb^{2+} 总量？为什么？

4. 滴定 Pb^{2+} 时要调节溶液 pH 为 5～6，为什么加入六亚甲基四胺而不加入醋酸钠？

● 视频演示

风险评估及控制工作表

实验任务: EDTA标准溶液的配制与标定及铅、铋混合液中铅、铋含量的连续测定

评估人: 吴泽颖、王红松、吴星星

评估日期: 2021年7月1日

过程步骤	化合物名称	危害鉴别				健康风险		降低或消除风险	安全措施			清单/风险评估	风险评估			紧急情况/泼洒	特殊备注
		类别	浓度	用量	化学品安全技术说明书	可能的伤害种类	或物理危害		技术	防护眼镜	手套	通风橱		可能性	危害	风险	危害级别	
A: 0.02 mol·L⁻¹ EDTA标准溶液的配制																		
	乙二胺四乙酸二钠盐	—	0.02 M	400 mL	√	皮肤/眼刺激		关注		√	√			1~4	1	1	高中低	大量水冲洗
B: 0.02 mol·L⁻¹ 锌标准溶液的配制																		
准确称取 $ZnSO_4·7H_2O$ 基准试剂 1.40~1.45 g 于 100 mL 小烧杯中,加入约一半的水溶解后,定量转入 250 mL 容量瓶中,稀释,定容,摇匀。	$ZnSO_4·7H_2O$	—	0.02 M	250 mL	√	眼损伤		关注		√	√			1	1	1	低	大量水冲洗
C: EDTA 溶液浓度的标定																		
1)准确移取 25.00 mL 锌标准溶液于锥形瓶中,加入 2.5 mL 1:5 HCl 溶液,二甲酚橙指示剂 2滴,滴加六亚甲基四胺溶液至溶液呈紫红色后,再过量 5 mL,摇匀。用待标定的 EDTA 溶液进行滴定,至溶液由紫红色恰好变为亮黄色为终点,记录 V_Y。平行标定 3 次。	六亚甲基四胺溶液	—	20%	25 mL	√	皮肤过敏		关注		√	√			1	1	1	低	大量水冲洗
	二甲酚橙指示剂	—	0.2%	1 mL	—	皮肤/眼刺激		关注		√	√			1	1	1	低	大量水冲洗

续表

风险评估及控制工作表

实验任务	EDTA 标准溶液的配制与标定及铝、铋混合液中铝、铋含量的连续测定				评估人	吴泽颖,王红松,吴星星				评估日期	2021 年 7 月 1 日		
过程步骤	危害鉴别				健康风险		安全措施			清单/风险评估	风险评估		措施
	化合物名称	类别	浓度	用量	化学品安全技术说明书	可能的伤害种类	降低或消除风险	技术	防护眼镜	手套	通风橱		可能性 危害 风险 危害级别	紧急情况/泼洒 特殊备注
					或物理危害								1~4 1~4	高中低

过程步骤	化合物名称	类别	浓度	用量	化学品安全技术说明书	可能的伤害种类	降低或消除风险	技术	防护眼镜	手套	通风橱	可能性	危害	风险	危害级别	紧急情况/泼洒
D: $Bi^{3+}-Pb^{2+}$ 的连续测定 用移液管移取 25.00 mL $Bi^{3+}-Pb^{2+}$ 混合液于 250 mL 锥形瓶中,加入 10 mL 0.1 mol·L^{-1} HNO_3 溶液,滴加 2 滴二甲酚橙,用 EDTA 滴定溶液由紫红色突变为亮黄色,即为终点,记录消耗 EDTA 的用量 V_1,滴加六亚甲基四胺溶液至溶液呈紫红色后,再过量 5 mL,此时溶液的 pH 应为 5~6,继续用 EDTA 滴定溶液由紫红色突变为亮黄色,记录消耗 EDTA 的用量 V_2。平行滴定 3 次。	HNO_3	8+ 5.1	0.1 M	10 mL	√	皮肤灼伤和眼损伤	关注		√	√		1	3	3	低	大量水冲洗
实验产物																
实验废弃物																
整个过程的最终评估	低风险(按照正确流程操作)															
签名	实验老师:						实验室主任/职业安全健康(OHS)代表:							学生/操作人:		

Experiment 16 Preparation and standardization of EDTA standard solution and determination of Bi^{3+} and Pb^{2+} in the mixed solution of bismuth and lead

【Objectives】

1. To master the method of preparation and standardization of standard EDTA solution.
2. To master the way to determine the end point of xylenol orange indicator.
3. To learn the principle of increasing the selectively of EDTA by controlling acidity.
4. To master the method of continuous titration using EDTA solution for various metal ions.

【Principles】

1. Preparation and standardization of EDTA standard solution

EDTA standard solution is still prepared by indirect method. In addition to the primary standard $CaCO_3$ used in experiment 15, $ZnSO_4 \cdot 7H_2O$ can also be used to titrate the EDTA standard solution. If $ZnSO_4 \cdot 7H_2O$ was used as the primary standard to titrate EDTA, xylenol orange (XO) was selected as the indicator and hydrochloric acid hexamethylenetetramine was used to adjust the pH of the solution to 5~6. The reaction of the end point is:

$$Zn-XO \text{ (purplish red)} + Y = ZnY + XO \text{ (yellow)}$$

2. Determination of Bi^{3+} and Pb^{2+} in the mixed solution of bismuth and lead

Bi^{3+} and Pb^{2+} can form stable 1 : 1 complexes with EDTA. $\lg K$ of them are 27.94 and 18.04, respectively. Because of the great different between the two values of $\lg K$, $\Delta \lg K > 6$, we can use the acid effect to titrate them separately by controlling different acidity.

In titration, with xylenol orange as indicator, adjust the acidity of solution to pH≈1.0 with HNO_3 and titrate Bi^{3+} with EDTA standard solution. Then adjust the acidity of solution to pH 5~6 with hexamethylenetetramine and continue to titrate Pb^{2+} with EDTA standard solution.

$$pH = 1.0: Bi^{3+} + XO \text{ (yellow)} = Bi-XO \text{ (purplish red)}$$
$$Bi^{3+} + Y = BiY$$
$$Bi-XO \text{ (purplish red)} + Y = BiY + XO \text{ (yellow)}$$
$$pH = 5\sim6: Pb^{2+} + XO \text{ (yellow)} = Pb-XO \text{ (purplish red)}$$
$$Pb^{2+} + Y = PbY$$
$$Pb-XO \text{ (purplish red)} + Y = PbY + XO \text{ (yellow)}$$

【Apparatus and Chemicals】

1. Apparatus

Alkali burette (50 mL); Volumetric flask (100 mL); Volumetric flask (250 mL); Volumetric flask (500 mL); Pipette (10 mL); Erlenmeyer flask (250 mL); Beaker (100 mL); Beaker (500 mL); Electronic balance.

2. Chemicals

Disodium ethylenediaminetetraacetate; $ZnSO_4 \cdot 7H_2O$; 0.2% xylenol orange solution; 20% hexamethylenetetramine; 0.1 mol·L^{-1} HNO_3 solution; 1:1 HCl solution; 1:5 HCl solution; mixed solution of Bi^{3+} and Pb^{2+} (the concentrations of Bi^{3+} and Pb^{2+} are 0.01 mol·L^{-1} respectively, and the concentrations of HNO_3 are about 0.15 mol·L^{-1}).

【Experimental Procedures】

1. Preparation of 0.02 mol·L^{-1} EDTA standard solution

Same as experiment 15.

2. Preparation of 0.02 mol·L^{-1} Zinc standard solution

Accurately weigh 1.40~1.45 g of $ZnSO_4 \cdot 7H_2O$ primary standard into a 100 mL beaker, add about half of the water to dissolve it. Transfer it into a 250 mL volumetric flask quantitatively, dilute it, fix the volume, and shake it well.

3. Standardization of EDTA standard solution

Accurately transfer 25.00 mL of zinc standard solution into an Erlenmeyer flask. Add 2.5 mL of 1:5 HCl solution and two drops of xylenol orange indicator. Add hexamethylenetetramine solution until the color of the solution is purplish red, then excess 5 mL, and shake well. Titrate with EDTA solution until the color of the solution changes from purplish red to bright yellow. Record the volume of the EDTA standard solution consumed in the titration (V_Y). Repeat 3 times.

4. Continuous titration of Bi^{3+} and Pb^{2+}

Transfer 25.00 mL of mixed solution of Bi^{3+} and Pb^{2+} into a 250 mL Erlenmeyer flask. Add 10 mL of 0.1 mol·L^{-1} HNO_3 solution and 2 drops of xylenol orange, the end point is when the color of the solution changes from purple red to bright yellow. Record he volume of the EDTA standard solution consumed in the titration (V_1). Add hexamethylenetetramine solution until the color of the solution is purple red, and then excess 5 mL. At this time, the pH of the solution should be 5~6. Continue to titrate until the color of the solution changes from purple red to bright yellow. Record the volume of the EDTA standard solution consumed in the titration (V_2). Repeat 3 times.

【Notes】

1. Due to the slow reaction speed between Bi^{3+} and EDTA, it should be added slowly near the end point and shake well.

2. Bi^{3+} is very easy to hydrolyze. The prepared mixed test solution must have a high concentration

of HNO_3. Dilute it to about 0.15 mol · L^{-1} with water before use. Before and at the beginning of titration, do not use more water to wash the conical bottleneck to prevent the hydrolysis of Bi^{3+}.

3. When the EDTA solution is calibrated and the content of Pb^{2+} is determined, it can be heated to 50~60 ℃, so that the end point can be easily identified. When measuring Pb^{2+} content, adjust pH after titration of Bi^{3+}, add EDTA solution to the near end point, heat the solution, then carefully titrate to the end point. Otherwise, white precipitate may appear, affecting the judgment of the end point.

【Questions】

1. In complexometric titration, what are the conditions for accurate separate titration?

2. According to this experiment, does the initial acidity of Bi^{3+} titration exceed the maximum acidity of Bi^{3+}? What is the acidity of the solution at the end of titration to Bi^{3+}? What is the pH of the solution after adding 10 mL 200 g · L^{-1} hexamethylenetetramine?

3. Can we take two equal parts of mixed test solution, one of which is used to titrate Bi^{3+} with pH≈1.0, and the other is used to titrate the total amount of Bi^{3+} and Pb^{2+} with pH 5~6? Why?

4. Why add hexamethylenetetramine instead of sodium acetate to adjust the pH of the solution to 5~6 to titrate Pb^{2+}?

Risk Assessment and Control Worksheet

TASK	Preparation and standardization of EDTA standard solution and Determination of Bi^{3+} and Pb^{2+} in the mixed solution of bismuth and lead					Assessor	Zeying WU, Hongsong WANG, Xingxing WU				Date	1 JUL. 2021			
Process Steps	Hazard Identification					Health Risk		Safety Measures				Checklist＼Risk Assessment			Action
	Chemical Name	Class or Physical Hazard	Concn.	Amount	MSDS	Type of Injury Possible	To reduce or eliminate the risk	Technique	Glasses	Gloves	Fume Cbd.	Likely	Hazard	Risk Score	Risk Level	In case of Emergency/Spill
												1～4	1～4		H M L	Special Comments
A: Preparation of 0.02 mol·L^{-1} EDTA standard solution																
	Ethylenediaminetetraacetic acid disodium salt	—	0.02 M	400 mL	✓	Skin and eye irritant	Handle with care		✓	✓		1	1	1	L	Sponge & water
B: Preparation of 0.02 mol·L^{-1} Zinc standard solution																
Accurately weigh 1.40～1.45 g of $ZnSO_4 \cdot 7H_2O$ primary standard into a 100 mL beaker, add about half of the water to dissolve it. Transfer it into a 250 mL volumetric flask quantitatively, dilute it, fix the volume, and shake it well.	$ZnSO_4 \cdot 7H_2O$	—	0.02 M	250 mL	✓	Eye injuries	Handle with care		✓	✓		1	1	1	L	Sponge & water
C: Standardization of EDTA standard solution																
Accurately transfer 25.00 mL of zinc standard solution into an Erlenmeyer flask. Add 2.5 mL of 1∶5 HCl solution and two drops of xylenol orange indicator. Add hexamethylenetetramine solution until the color of the solution is purplish red, then excess 5 mL, and shake well. Titrate with EDTA solution until the color of the solution changes from purplish red to bright yellow. Record the volume of the EDTA standard solution consumed in the titration (V_Y). Repeat 3 times.	Hexamethylenetetramine solution	—	20%	25 mL	✓	Skin allergy	Handle with care		✓	✓		1	1	1	L	Sponge & water
	Xglenol orange indicator	—	0.2%	1 mL	—	Skin and eye irritant	Handle with care		✓	✓		1	1	1	L	Sponge & water

continued

Risk Assessment and Control Worksheet

TASK	Process Steps	Assessor	Date					
Preparation and standardization of EDTA standard solution and Determination of Bi^{3+} and Pb^{2+} in the mixed solution of bismuth and lead		Zeying WU, Hongsong WANG, Xingxing WU	1 JUL. 2021					

Hazard Identification				Health Risk	Safety Measures				Checklist\Risk Assessment	Risk Assessment			Action	
Chemical Name	Class or Physical Hazard	Concn.	Amount	MSDS	Type of Injury Possible	To reduce or eliminate the risk	Technique	Glasses	Gloves	Fume Cbd.	Likely 1~4	Hazard 1~4	Risk Score	Risk Level H M L	In case of Emergency/Spill	Special Comments
HNO_3		8+5.1	0.1 M	10 mL	✓	Skin burns and eye injuries	Handle with care		✓	✓		1	3	3	L	Sponge & water

D: Continuous titration of Bi^{3+} and Pb^{2+}

Transfer 25.00 mL of mixed solution of Bi^{3+} and Pb^{2+} into a 250 mL Erlenmeyer flask. Add 10 mL of 0.1 mol·L^{-1} HNO_3 solution and 2 drops of xylenol orange, the end point is when the color of the solution changes from purple red to bright yellow. Record he volume of the EDTA standard solution consumed in the titration (V_1). Add hexamethylenetetramine solution until the color of the solution is purple red, and then excess 5 mL. At this time, the pH of the solution should be 5~6. Continue to titrate until the color of the solution changes from purple red to bright yellow. Record the volume of the EDTA standard solution consumed in the titration (V_2). Repeat 3 times.

continued

Risk Assessment and Control Worksheet

TASK	Preparation and standardization of EDTA standard solution and Determination of Bi^{3+} and Pb^{2+} in the mixed solution of bismuth and lead				Assessor	Zeying WU, Hongsong WANG, Xingxing WU				Date	1 JUL. 2021			
Process Steps	Hazard Identification				Health Risk	Safety Measures				Checklist＼Risk Assessment	Risk Assessment			Action
	Chemical Name	Class or Physical Hazard	Concn.	Amount	MSDS	Type of Injury Possible	To reduce or eliminate the risk	Technique	Glasses	Gloves	Fume Cbd.		Likely	Hazard	Risk Score	Risk Level	In case of Emergency/Spill
													1~4	1~4		H M L	Special Comments
Experimental Product(s)																	
Experimental Waste(s)																	
Final assessment of overall process	Low risk when correct procedures followed																
Signatures	Supervisor:					Lab Manager/OHS Representative:										Student/Operator:	

实验 17　高锰酸钾标准溶液的配制与浓度标定及过氧化氢的含量测定

【实验目的】

1. 了解高锰酸钾（$KMnO_4$）溶液的保存条件，掌握其标准溶液的配制方法。
2. 掌握用无水 $Na_2C_2O_4$ 作为基准物质标定 $KMnO_4$ 溶液的基本原理、操作方法、滴定条件和计算。
3. 了解自动催化反应的特点。
4. 掌握过氧化氢试样的称量方法。
5. 掌握应用高锰酸钾直接滴定法测定双氧水中过氧化氢含量的基本原理、操作方法和计算。

【实验原理】

1. 高锰酸钾溶液的配制——间接法

首先，高锰酸钾溶液不能用直接法配制的原因有以下几点：

(1) 市售 $KMnO_4$ 常含有少量 MnO_2 和其他杂质；

(2) 蒸馏水中含有少量有机物质。

它们能使 $KMnO_4$ 还原为 $MnO(OH)_2$，而 $MnO(OH)_2$ 又能促进 $KMnO_4$ 自身的分解：$4MnO_4^- + 2H_2O =\!=\!= 4MnO_2 + 3O_2\uparrow + 4OH^-$，见光时分解得更快。

2. 高锰酸钾溶液的保存

(1) 中性介质；

(2) 不含 MnO_2 等杂质；

(3) 保存于暗处；

(4) 如长期使用，仍应定期标定。

3. 高锰酸钾溶液的标定（"三度一点"）

(1) 温度：75～85 ℃。温度过高，在酸性溶液中部分 $H_2C_2O_4$ 发生分解：

$$H_2C_2O_4 =\!=\!= CO_2\uparrow + CO\uparrow + H_2O$$

(2) 酸度，H_2SO_4 的浓度约为 0.5～1 mol·L^{-1}。酸度低时，易生成 MnO_2 沉淀；酸度过高，加速 $H_2C_2O_4$ 分解。

(3) 滴定速度，慢—快—慢（自动催化反应）。待第一滴 $KMnO_4$ 红色褪去后，再滴入第二滴。否则，加入的 $KMnO_4$ 溶液来不及反应，在热的酸性溶液中发生分解：$4MnO_4^- + 12H^+ =\!=\!= 4Mn^{2+} + 5O_2\uparrow + 6H_2O$

(4) 滴定终点时溶液由无色变为粉红色，且 30 s 不褪色。由于空气中的还原性气体及尘埃等杂质落入溶液中能使 $KMnO_4$ 缓慢分解，而使粉红色消失，所以经过 30 s 不褪色即为终点。

在酸度为 0.5 mol·L^{-1} 的 H_2SO_4 酸性溶液中，以 $Na_2C_2O_4$ 为基准物质标定高锰酸钾溶液，以高锰

酸钾自身为指示剂,反应式为:

$$2MnO_4^- + 5C_2O_4^{2-} + 16H^+ =\!\!=\!\!= 2Mn^{2+} + 10CO_2\uparrow + 8H_2O$$

4. 双氧水中过氧化氢含量的测定

在酸性溶液中 H_2O_2 是强氧化剂,但遇到强氧化剂 $KMnO_4$ 时,又表现为还原剂。因此,可以在酸性溶液中用 $KMnO_4$ 标准溶液直接滴定测得 H_2O_2 的含量,以 $KMnO_4$ 自身为指示剂。反应式为:

$$2MnO_4^- + 5H_2O_2 + 6H^+ =\!\!=\!\!= 2Mn^{2+} + 5O_2\uparrow + 8H_2O$$

根据 $M(1/2\ H_2O_2)$ 和 $C(1/5\ KMnO_4)$ 以及滴定中消耗 $KMnO_4$ 的体积计算 H_2O_2 的含量。

【器材与试剂】

1. 器材:250 mL 锥形瓶,50 mL 酸式滴定管,250 mL 容量瓶,25.00 mL 移液管,10 mL 量筒,500 mL 量筒,500 mL 烧杯,称量瓶,干燥器,分析天平,加热板。

2. 试剂:$KMnO_4$ 固体,$Na_2C_2O_4$ 基准物质(在 105~110 ℃烘干至恒重),3 mol·L^{-1} H_2SO_4 溶液、30% H_2O_2 试样。

【实验内容】

1. 0.02 mol·L^{-1} 的 $KMnO_4$ 溶液配制 400 mL

称取约 1.3 g $KMnO_4$ 固体于烧杯中,加 400 mL 去离子水使之完全溶解,加热煮沸 20~30 min(随时加水以补充因蒸发而损失的水),冷却后于暗处静置 2~3 天,随后用玻璃砂芯漏斗过滤除去 MnO_2 等杂质,滤液贮存于洁净的棕色试剂瓶中。若溶液煮沸后在水浴上保温 1 h,则不必长期放置,可直接冷却后过滤,待标定。

2. $KMnO_4$ 溶液的标定

准确称取 0.15~0.20 g $Na_2C_2O_4$ 基准物质,置于 250 mL 锥形瓶中,加入 30 mL 水使之溶解,再加入 10 mL 3 mol·L^{-1} H_2SO_4 溶液,加热至 75~85 ℃,立即用待标定的 $KMnO_4$ 溶液滴定(不能沿瓶壁滴入)直至呈粉红色,30 s 内不褪色即为终点。记录消耗的 $KMnO_4$ 溶液体积,平行测定四次,同时做空白实验。

3. H_2O_2 含量的测定

准确移取 H_2O_2 试样 2.00 mL 于 250 mL 容量瓶中,定容。移取 25.00 mL 此溶液置于 250 mL 锥形瓶中,加入 10~20 mL 去离子水和 10 mL 3 mol·L^{-1} H_2SO_4 溶液,再用 0.02 mol·L^{-1} 的 $KMnO_4$ 标准溶液滴定至溶液呈粉红色,30 s 内不褪色,即为终点。记录消耗的 $KMnO_4$ 溶液体积,平行测定四次,同时做空白实验。

【注意事项】

1. 分析天平的正确使用;有色溶液液面的观察与正确读数。
2. $KMnO_4$ 溶液配制过程中,需要用玻璃砂芯漏斗过滤,滤液贮存于棕色瓶中,放置于暗处保存。
3. 双氧水有腐蚀性,使用时需带手套;应在室温下配制试液,且定容静置前要将试液摇匀。
4. 指示剂终点颜色的观察。

【思考题】

1. 配制高锰酸钾溶液时，为什么要用玻璃砂芯漏斗过滤？能否直接使用常规的定量滤纸过滤？为什么？

2. 配制高锰酸钾溶液时，为什么煮沸冷却后的高锰酸钾溶液要静置几天才能使用？

3. 高锰酸钾法测定 H_2O_2 含量为什么要在酸性溶液中进行？为什么选用的是 H_2SO_4 而不是 HNO_3 或者 HCl 等？

4. H_2O_2 含量的测定过程中，为什么不直接移取 0.2 mL 30% 双氧水进行测定，而是要稀释一定倍数后再移取？

5. H_2O_2 与 $KMnO_4$ 反应较慢，能否通过加热溶液的方式来加快反应速率？

● 视频演示

风险评估及控制工作表

实验任务	高锰酸钾标准溶液的配制与浓度标定及过氧化氢的含量测定			评估人	吴泽颖、王红松、吴星星				评估日期	2021 年 7 月 1 日			
过程步骤	危害鉴别			健康风险		安全措施			清单/风险评估	风险评估			紧急情况/泼洒特殊备注
	类别	浓度	用量	可能的伤害种类	化学品安全技术说明书	降低或消除风险	技术	防护眼镜	手套	通风橱	可能性 1~4	危害 1~4	风险	危害级别 高中低	措施
化合物名称	或物理危害														

A: $KMnO_4$ 溶液配制

称取约 1.3 g $KMnO_4$ (Mr = 158.0) 固体于烧杯中,加 400 mL 去离子水使之完全溶解,加热煮沸 20~30 min(随时加水以补充因蒸发而损失的水),冷却后于暗处静置 2~3 天。随后用玻璃砂芯漏斗过滤除去 MnO_2 等杂质,滤液贮存于洁净的棕色试剂瓶中。若溶液煮沸后在水浴上保温 1 h,则不必长期放置,可直接冷却后过滤,待标定。

| $KMnO_4$ | 5.1 | 0.02 M | 400 mL | — | √ | 关注 | | √ | √ | | 1 | 1 | 1 | 低 | 大量水冲洗 |
| 电加热装置 | | | | 电击 & 烫伤 | | 确保电线已贴上标签 | √ | √ | √ | | 1 | 3 | 3 | 低 | 立即向实验老师或准备实验人员报告 |

B: $KMnO_4$ 溶液的标定

准确称取 0.15~0.20 g $Na_2C_2O_4$ 基准物质,置于 250 mL 锥形瓶中,加入 30 mL 水使之溶解,再加入 10 mL 3 mol·L^{-1} H_2SO_4 溶液,加热至 75~85 ℃,立即用待标定的 $KMnO_4$ 溶液滴定(不能沿瓶壁滴入)直至呈粉红色,30 s 内不褪色即为终点。记录消耗的 $KMnO_4$ 溶液体积,平行测定四次,同时做空白实验。

| $Na_2C_2O_4$ | 4 | — | 0.2 g | 吞咽和皮肤接触有害 | √ | 关注 | | √ | √ | | 1 | 3 | 3 | 低 | 大量水冲洗 |
| H_2SO_4 | 8 | 3 M | 10 mL | 皮肤灼伤和眼损伤 | √ | 关注 | | √ | √ | | 1 | 4 | 4 | 低 | 大量水冲洗 |

续表

风险评估及安控制工作表

实验任务	高锰酸钾标准溶液的配制与浓度标定及过氧化氢的含量测定			评估人	吴泽颖、王红松、吴星星				评估日期	2021年7月1日		
过程步骤	危害鉴别			健康风险	安全措施				清单/风险评估	风险评估		紧急情况/泄洒措施
	类别	浓度	用量	化学品安全技术说明书	可能的伤害种类	降低或消除风险	防护眼镜	技术	手套	通风橱	可能性 1~4	危害 1~4	风险 3	危害级别 高中低

化合物名称 | 类别 | 浓度 | 用量 | 化学品安全技术说明书 | 可能的伤害种类 | 降低或消除风险 | 防护眼镜 | 技术 | 手套 | 通风橱 | 可能性 | 危害 | 风险 | 危害级别 | 紧急情况/泄洒措施/特殊备注

或物理危害

化合物名称	类别	浓度	用量	化学品安全技术说明书	可能的伤害种类	降低或消除风险	防护眼镜	手套	通风橱	可能性	危害	风险	危害级别	紧急情况/泄洒 特殊备注
H_2O_2	5.1 +8	30%	250 mL	√	皮肤灼伤和眼损伤	关注	√	√		1	3	3	低	大量水冲洗

C: H_2O_2 含量的测定

准确移取 H_2O_2 试样 5.0 mL 于 250 mL 容量瓶中,定容。移取 10.0 mL 此溶液置于 250 mL 锥形瓶中,加入 10~20 mL 去离子水和 10 mL 3 mol·L^{-1} H_2SO_4 溶液,再用 0.02 mol·L^{-1} 的 $KMnO_4$ 标准溶液滴定至溶液呈粉红色,30 s 内不褪色,即为终点。记录消耗的 $KMnO_4$ 溶液体积,平行测定四次,同时做空白实验。

实验产物	
实验废弃物	
整个过程的最终评估	低风险(按照正确流程操作)
签名	实验老师: 实验室主任/职业安全健康(OHS)代表: 学生/操作人:

Experiment 17　Preparation and calibration of potassium permanganate standard solution and determination of hydrogen peroxide content

【Objectives】

1. To understand the storage conditions of potassium permanganate ($KMnO_4$) solution, and master the preparation method of its standard solution.

2. To grasp the basic principle, operation method, titration condition and calculation of calibrating $KMnO_4$ solution with anhydrous $Na_2C_2O_4$ as the reference material.

3. To know the characteristics of autocatalytic reactions.

4. To master the weighing method of hydrogen peroxide sample.

5. To acquire the basic principle, operation and calculation method of hydrogen peroxide content determination directly by potassium permanganate titration.

【Principles】

1. Preparation of $KMnO_4$ solution——indirect method

Firstly, $KMnO_4$ solution cannot be configured directly for the following reasons:

(1) There are always small amounts of MnO_2 and other impurities in the commercial $KMnO_4$.

(2) A small amount of organic matter is contained in the distilled water, which can lead to $KMnO_4$ being reduced to $MnO(OH)_2$. Furthermore, $MnO(OH)_2$ can facilitate the self-decomposition of $KMnO_4$: $4MnO_4^- + 2H_2O = 4MnO_2 + 3O_2\uparrow + 4OH^-$. And this decomposition process will go faster with light.

2. Storage of $KMnO_4$ solution

(1) In the neutral medium.

(2) Free of MnO_2 and other impurities.

(3) Keeping in the dark.

(4) Being calibrated regularly if used for a long time.

3. Calibration of $KMnO_4$ solution

(1) Temperature: 75~85 ℃. If the temperature is too high, part of $H_2C_2O_4$ will decompose in the acidic solution: $H_2C_2O_4 = CO_2\uparrow + CO\uparrow + H_2O$

(2) Acidity: the concentration of H_2SO_4 should be 0.5~1 mol·L^{-1}. If the acidity is too low, it is easy to form MnO_2 precipitation. On the contrary, it will accelerate $H_2C_2O_4$ decomposition.

(3) The speed of titration: slow to fast, and then to slow. It is a self-decomposition process.

After the first drop of red $KMnO_4$ has faded, a second drop is added. Or, it will be too late to react for the added $KMnO_4$, which then decompose in the hot and acid solution: $4MnO_4^- + 12H^+ \mathrm{=\!=\!=} 4Mn^{2+} + 5O_2\uparrow + 6H_2O$.

(4) The terminal point of titration: colourless to pink. Because of the impurities such as reducing gas and dust in the air falling into the solution, $KMnO_4$ can be decomposed slowly, resulting in the pink disappearing. Therefore, the titration end can be indicated by the red $KMnO_4$ not fading after 30 s.

In H_2SO_4 solution (0.5~1 mol·L^{-1}), potassium permanganate solution is calibrated with $Na_2C_2O_4$ as the reference and $KMnO_4$ itself as the indicator. The reaction formula is as follow: $2MnO_4^- + 5C_2O_4^{2-} + 16H^+ \mathrm{=\!=\!=} 2Mn^{2+} + 10CO_2\uparrow + 8H_2O$

4. Determination of H_2O_2 in hydrogen peroxide solution

H_2O_2, a strong oxidizer in acidic solution, can act as a reducing agent when it encounters $KMnO_4$ that is a strong oxidizer. So, H_2O_2 content can be determined by directly being titrated with $KMnO_4$ standard solution in acidic solution, during which $KMnO_4$ itself is the indicator. The reaction formula is as follow: $2MnO_4^- + 5H_2O_2 + 6H^+ \mathrm{=\!=\!=} 2Mn^{2+} + 5O_2\uparrow + 8H_2O$.

According to $M(1/2\ H_2O_2)$, $C(1/5\ KMnO_4)$ and the consumed volume of $KMnO_4$, the content of H_2O_2 in the commercial hydrogen peroxide solution can be calculated.

【Apparatus and Chemicals】

1. Apparatus

Conical bottle (250 mL); acid burette (50 mL); volumetric bottle (250 mL); pipette (25.00 mL); measuring cylinder (500 mL); measuring cylinder (10 mL); beaker (500 mL); weighing bottle; dryer; analytical balance, hotplate.

2. Chemicals

$KMnO_4$ solid; $Na_2C_2O_4$ primary standard substance (being dried at 105~110 ℃ to constant weight); H_2SO_4 solution (3 mol·L^{-1}); 30% H_2O_2 solution.

【Experimental Procedures】

1. Preparation of 400 mL 0.02 mol·L^{-1} $KMnO_4$ solution

About 1.3 g $KMnO_4$ solid is weighed and put in a beaker, with 400 mL deionized water added to completely dissolve it. Then the obtained solution is heated and kept boiling state for 20~30 min (supplementing water persistently that has been being evaporated). After cooled and stood in the dark for 2~3 days, it was filtered with a glass sand core funnel to remove impurities such as MnO_2. The filtrate was stored in a clean brown reagent bottle. If the solution is kept in the water bath for 1 h after being boiled, there is no need for it being placed for a long time. So, the treated solution can be directly cooled and filtered for calibration.

2. Calibration of $KMnO_4$ solution

0.15~0.20 g $Na_2C_2O_4$ (reference material) should be firstly accurately weighed before being put into a 250 mL conical bottle. After 30 mL water is added and it is completely dissolved, 10 mL

3 mol·L^{-1} H$_2$SO$_4$ solution is added and the resulted solution is then heated to 75 ~ 85 ℃. Subsequently, titrating it immediately with KMnO$_4$ solution that is to be calibrated (not dripping along the wall of the bottle) until the solution becomes pink. If there is no colour fading within 30 s, the titration end point comes. The consumed volume of KMnO$_4$ solution should be recorded and the parallel tests should be carried out four times. Blank experiment is also necessary.

3. Determination of H$_2$O$_2$ content

Accurately transfer 2.00 mL H$_2$O$_2$ sample to a 250 mL volumetric flask, dilute with deionized water and mix. Then, transfer 25.00 mL of the resulted solution to a 250 mL conical flask, add 10 ~ 20 mL deionized water and 10 mL H$_2$SO$_4$ (3 mol·L^{-1}) solution, followed by titrating with 0.02 mol·L^{-1} of KMnO$_4$ standard solution until the solution is pink and does not fade within 30 s, which is the end point. Record the volume of KMnO$_4$ solution consumed, measure it four times in parallel, and do blank experiment at the same time.

【Notes】

1. Proper use of analytical balance; Observation and correct reading of colored solution.
2. During the KMnO$_4$ solution configuration, it is necessary to use glass sand core funnel to filter. And the obtained filtrate shall be stored in brown bottles and in the dark.
3. Gloves are required because hydrogen peroxide is corrosive. The test solution should be prepared at room temperature, and should be shaken well before standing at the constant volume.
4. Careful observation of the indicator end color.

【Questions】

1. Why using glass sand core funnel to filter potassium permanganate solution? Can we use regular quantitative filter directly? Why or not?
2. Why is potassium permanganate solution placed in the dark for several days after being boiled and cooled before use?
3. Why is the determination of H$_2$O$_2$ content performed in acidic solution? Why is H$_2$SO$_4$ choosed? Is HNO$_3$ or HCl ok?
4. During the determination of H$_2$O$_2$ content, why not directly remove 0.2 mL 30% hydrogen peroxide? What is the reason of diluting the origin H$_2$O$_2$ solution by a certain number of times before use?
5. Considering the fact that the reaction of H$_2$O$_2$ and KMnO$_4$, can we speed up the reaction by direct heating the mixed solution?

Risk Assessment and Control Worksheet

TASK	Assessor	Date
Preparation and concentration calibration of potassium permanganate standard solution and content determination of hydrogen peroxide	Zeying WU, Hongsong WANG, Xingxing WU	1 JUL. 2021

Process Steps	Hazard Identification				Health Risk		Safety Measures				Checklist~Risk Assessment				Action	
	Chemical Name	Class or Physical Hazard	Concn.	Amount	MSDS	Type of Injury Possible	To reduce or eliminate the risk	Technique	Glasses	Gloves	Fume Cbd.	Likely 1~4	Hazard 1~4	Risk Score	Risk Level H M L	In case of Emergency/Spill	Special Comments
	KMnO$_4$	5.1	0.02 M	400 mL	√	—	Handle with care			√		1	1	1	L	Sponge & water	
	Heating device					Electric shock & scald	Ensure cord is tagged	√	√			1	3	3	L	Report immediately to Demonstrator or Prep Room	

A: Preparation of 400 mL 0.02 mol·L^{-1} KMnO$_4$ solution

About 1.3 g KMnO$_4$ (M_r = 158.0) solid is weighed and put in a beaker, with 400 mL deionized water added to completely dissolve it. Then the obtained solution is heated and kept boiling state for 20~30 min (supplementing water persistently that has been being evaporated). After cooled and stood in the dark for 2~3 days, it was filtered with a glass sand core funnel to remove impurities such as MnO$_2$. The filtrate was stored in a clean brown reagent bottle. If the solution is kept in the water bath for 1 h after being boiled, there is no need for it being placed for a long time. So, the treated solution can be directly cooled and filtered for calibration.

Risk Assessment and Control Worksheet continued

TASK	Preparation and concentration calibration of potassium permanganate standard solution and content determination of hydrogen peroxide					Assessor	Zeying WU, Hongsong WANG, Xingxing WU					Date	1 JUL. 2021		
Process Steps	Hazard Identification				Health Risk	Safety Measures					Checklist～Risk Assessment		Risk Assessment		Action
	Chemical Name	Class	Concn.	Amount	MSDS	Type of Injury Possible	To reduce or eliminate the risk	Technique	Glasses	Gloves	Fume Cbd.	Likely 1～4	Hazard 1～4	Risk Score	Risk Level H M L	In case of Emergency/Spill Special Comments

B: Calibration of KMnO₄ solution

0.15～0.20 g Na₂C₂O₄ (reference material) should be firstly accurately weighed before being put into a 250 mL conical bottle. After 30 mL water is added and it is completely dissolved, 10 mL 3 mol·L⁻¹ H₂SO₄ solution is added and the resulted solution is then heated to 75～85 ℃. Subsequently, titrating it immediately with KMnO₄ solution that is to be calibrated (not dripping along the wall of the bottle) until the solution becomes pink. If there is no colour fading within 30 s, the titration end point comes. The consumed volume of KMnO₄ solution should be recorded and the parallel tests should be carried out four times. Blank experiment is also necessary.	Na₂C₂O₄	4		0.2 g	√	Toxic in contact with skin or swallowed	Handle with care		√	√		1	3	3	L	Sponge & water
	H₂SO₄	8	3 M	10 mL	√	Skin burns and eye injuries	Handle with care		√	√		1	4	4	L	Sponge & water

continued

Risk Assessment and Control Worksheet

TASK	Preparation and concentration calibration of potassium permanganate standard solution and content determination of hydrogen peroxide					Assessor	Zeying WU, Hongsong WANG, Xingxing WU					Date	1 JUL. 2021		
Process Steps	Hazard Identification					Health Risk	Safety Measures					Checklist～Risk Assessment	Risk Assessment		Action
	Chemical Name	Class	Concn.	Amount	MSDS	Type of Injury Possible	To reduce or eliminate the risk	Technique	Glasses	Gloves	Fume Cbd.	Likely / Hazard	Risk Score	Risk Level	In case of Emergency/Spill
		or Physical Hazard										1～4 / 1～4		H M L	Special Comments
C: Determination of H₂O₂ content															
Accurately transfer 5.0 mL H$_2$O$_2$ sample to a 250 mL volumetric flask, dilute with deionized water and mix. Then, transfer 10.0 mL of the resulted solution to a 250 mL conical flask, add 10～20 mL deionized water and 10 mL H$_2$SO$_4$ (3 mol·L^{-1}) solution, followed by titrating with 0.02 mol·L^{-1} of KMnO$_4$ standard solution until the solution is pink and does not fade within 30 s, which is the end point. Record the volume of KMnO$_4$ solution consumed, measure it four times in parallel, and do blank experiment at the same time.	H$_2$O$_2$	5.1+8	30%	250 mL	√	Skin burns and eye injuries	Handle with care		√	√		1 / 3	3	L	Sponge & water
Experimental Product (s)															
Experimental Waste (s)															
Final assessment of overall process						Low risk when correct procedures followed									
Signatures	Supervisor:						Lab Manager/OHS Representative:						Student/Operator:		

实验 18　铁矿石中铁含量的测定（无汞法）

【实验目的】

1. 掌握无汞法测定铁含量的基本原理及方法。
2. 学会直接配制 $K_2Cr_2O_7$ 标准溶液的方法及其用于铁含量测定的原理、方法和实验条件，进一步地了解 $K_2Cr_2O_7$ 法在实际分析中的应用。
3. 了解酸分解铁矿石试样的方法及氧化还原的预处理过程，明确预处理的重要性。
4. 学习氧化还原指示剂的变色原理。
5. 增强环保意识。

【实验原理】

铁矿石的种类有磁铁矿（Fe_3O_4）、赤铁矿（Fe_2O_3）、菱铁矿（$FeCO_3$）等，铁矿石中铁含量的测定，一般采用经典的 $K_2Cr_2O_7$ 分析法，简便、快速。但因 $HgCl_2$ 含剧毒且污染环境，因此，为了减少实验废水对环境的污染，本实验采用改进的重铬酸钾法，在保留 $K_2Cr_2O_7$ 作为滴定剂的前提条件下，不使用 $HgCl_2$。

1. 铁矿石预处理

铁矿石样品在热的浓 HCl 和 $SnCl_2$ 溶液中溶解后，趁热用 $SnCl_2$ 溶液将大部分 Fe(Ⅲ) 还原为 Fe(Ⅱ)，试液由棕红色变为浅黄色。再以钨酸钠为指示剂，用 $TiCl_3$ 还原剩余的 Fe(Ⅲ) 至出现蓝色的 W(Ⅴ)（俗称钨蓝），即表明 Fe(Ⅲ) 已被全部还原，其反应式为：

$$SnCl_4^{2-} + 2Fe^{3+} + 2Cl^- = SnCl_6^{2-} + 2Fe^{2+}$$

$$Fe^{3+} + Ti^{3+} + H_2O = TiO^{2+} + Fe^{2+} + 2H^+$$

用少量稀的 $K_2Cr_2O_7$ 溶液滴至上述钨蓝刚好褪去，从而指示预还原的终点。

2. $K_2Cr_2O_7$ 滴定 Fe^{2+}

上述预处理后的溶液在硫磷混酸介质中，以二苯胺磺酸钠为指示剂，用 $K_2Cr_2O_7$ 标准溶液滴定至溶液呈紫色，即达滴定终点，涉及的反应式为：

$$Cr_2O_7^{2-} + 6Fe^{2+} + 14H^+ = 2Cr^{3+} + 6Fe^{3+} + 7H_2O$$

所以 Fe 的百分含量可以按下式计算：

$$w = \frac{m_{Fe}}{G_{试样}} \times 100\% = \frac{6C_{Cr_2O_7^{2-}} \cdot V_{Cr_2O_7^{2-}} \times \frac{M_{Fe}}{1\,000}}{G_{试样}} \times 100\%$$

值得注意的是，上述过程中，滴定之所以要在硫磷混酸介质中进行，主要是因为：

（1）为了使铁电（Fe^{3+}/Fe^{2+}）的电极电位降低变负，滴定突跃范围变宽，指示剂的变色点落在突跃范围内,提高滴定的准确度(选用二苯胺磺酸钠作为指示剂,滴定终点为 0.85 V,落在 0.86～1.26 之外)。

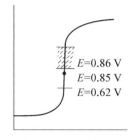

图 18-1　滴定曲线示意图

(2) 消除 Fe^{3+} 对终点观察的干扰。

【器材与试剂】

1. 器材:50 mL 酸式滴定管,250 mL 锥形瓶,250 mL 容量瓶,50 mL 量筒,500 mL 烧杯,100 mL 烧杯,称量瓶,表面皿,干燥器,分析天平,酒精灯。

2. 试剂:铁矿石固体,$K_2Cr_2O_7$ 基准试剂(在 140 ℃干燥 2 h),10% $SnCl_2$ 水溶液,$TiCl_3$ 溶液,HCl 溶液(6 mol·L^{-1}),硫-磷混合酸,25% Na_2WO_4 水溶液,0.2%二苯胺磺酸钠水溶液。

【实验内容】

1. 0.017 mol·L^{-1} 的 $K_2Cr_2O_7$ 溶液配制

称取 1.2～1.3 g $K_2Cr_2O_7$ 基准试剂于 100 mL 烧杯中,加适量水溶解后转入 250 mL 容量瓶中,定容。

2. 铁矿石试样的分解

准确称取 0.15～0.20 g 铁矿石样品,置于 250 mL 锥形瓶中并用少量水湿润,加入 20 mL HCl,滴加 8～10 滴 $SnCl_2$ 溶液(助溶),摇晃锥形瓶 1～2 min 以分散样品,随后盖上表面皿,在近沸的水中加热 20～30 min,至试样完全分解,残渣变为白色,此时混合溶液呈橙黄色。用少量水吹洗表面皿和锥形瓶内壁。

3. 铁矿石样品的预处理

趁热向上述锥形瓶中的溶液滴加 $SnCl_2$ 溶液以还原 Fe^{3+} 直至溶液由棕黄色变成浅黄色。再滴加 4 滴 Na_2WO_4 溶液和 50 mL H_2O,加热并摇动锥形瓶,同时向瓶内滴加 $TiCl_3$ 直至溶液呈稳定的浅蓝色。用水冲洗瓶壁,快速冷却至室温,小心滴加 2～5 滴 $K_2Cr_2O_7$ 溶液至蓝色刚好消失。

4. 铁含量的测定

向上述预处理过的溶液中加水稀释至 150 mL,再加 15 mL H_2SO_4-H_3PO_4 混合酸溶液,随后加入 5～6 滴二苯胺磺酸钠指示剂,并立即用 $K_2Cr_2O_7$ 溶液滴定至溶液呈稳定的紫色即为终点。记录消耗的 $K_2Cr_2O_7$ 溶液体积,并做 3 组平行实验。

【注意事项】

1. 铁矿石的分解过程需在低温下进行,因为 $FeCl_3$ 的沸点为 315 ℃且在 100 ℃左右就会显著挥发。
2. 分解完全的试样应只剩 SiO_2,为白色,否则即为分解不完全。

3. 为防止 Fe^{2+} 被氧化,预还原和滴定的溶液要各准备一瓶,且不要将两瓶溶液同时预还原。

4. 定量还原 Fe^{3+} 时,不能单独用 $SnCl_2$ 或者 $TiCl_3$,只能采用 $SnCl_2$-$TiCl_3$ 联合预还原法。

5. $SnCl_2$ 还原 Fe^{3+} 时,溶液体积不宜太大,以免影响对 Fe^{3+} 颜色的观察。另外,温度也不能低于 60 ℃,否则会导致反应进行缓慢,造成 $SnCl_2$ 过量不易除去。

6. 滴定过程中二苯胺磺酸钠指示剂会消耗一定量的 $K_2Cr_2O_7$ 溶液,故不能加太多。

【思考题】

1. 定量还原 Fe^{3+} 时,为什么不能单独用 $SnCl_2$ 或者 $TiCl_3$,而是要采用 $SnCl_2$-$TiCl_3$ 联合预还原法?只使用其中一种有什么弊端?

2. 在滴定前加入磷酸的作用是什么?加入磷酸后为什么要立即滴定?

● 配套课件

风险评估及控制工作表																	
实验任务	铁矿石中铁含量的测定(无汞法)					评估人	吴泽颖、王红松、吴星星					评估日期	2021 年 7 月 1 日				
过程步骤	危害鉴别					健康风险	安全措施					清单/风险评估	风险评估				措施
	化合物名称	类别	浓度	用量	化学品安全技术说明书	可能的伤害种类	降低或消除风险	技术	防护眼镜	手套	通风橱		可能性	危害	风险	危害级别	紧急情况/泼洒
	或物理危害												1～4	1～4		高中低	特殊备注
A:0.017 mol・L^{-1}的 $K_2Cr_2O_7$ 溶液配制	$K_2Cr_2O_7$	9	0.017 M	250 mL	√	皮肤灼伤和眼损伤	关注		√	√			1	3	3	低	大量水冲洗
B:铁矿石试样的分解																	
准确称取 0.15～0.20 g 铁矿石样品,置于 250 mL 锥形瓶中并用少量水湿润,加入 20 mLHCl,滴加 8～10 滴 $SnCl_2$溶液(助溶),摇晃锥形瓶 1～2 min 以分散样品,随后盖上表面皿,在近沸的水中加热 20～30 min,至试样完全分解,残渣变为白色,此时混合溶液呈橙黄色。用少量水吹洗表面皿和锥形瓶内壁。	HCl 溶液	2.3 +8	6 M	20 mL	√	皮肤灼伤和眼损伤	关注		√	√	√		1	4	4	低	大量水冲洗
	$SnCl_2$	溶液	10%	1 mL	√	皮肤烧伤和眼损伤	关注		√	√			1	3	3	低	大量水冲洗
	酒精灯	8	—	—		烧伤	关注	√	√				1	3	3	低	立即向指导老师或准备实验人员报告
C:铁矿石样品的预处理																	
趁热向上述锥形瓶中的溶液滴加 $SnCl_2$溶液以还原 Fe^{3+} 直至溶液由棕黄色变成浅黄色。再滴加 4 滴 Na_2WO_4溶液和 50 mLH_2O,加热并摇动锥形瓶,同时向瓶内滴加 $TiCl_3$直至溶液呈稳定的浅蓝色。用水冲洗瓶壁,快速冷却至室温,小心滴加 2～5 滴 $K_2Cr_2O_7$溶液至蓝色刚好消失。	$TiCl_3$	—	15%～20%	—	√	皮肤灼伤和眼损伤	关注		√	√			1	3	3	低	大量水冲洗

续表

风险评估及控制工作表

实验任务	铁矿石中铁含量的测定（无汞法）				评估人	吴泽颖,王红松,吴星星				评估日期	2021 年 7 月 1 日			
过程步骤	危害鉴别				健康风险		安全措施			清单/风险评估	风险评估			措施
	化合物名称	类别	浓度	用量	化学品安全技术说明书	可能的伤害种类	降低或消除风险	技术	防护眼镜	手套	通风橱		可能性	危害	风险	危害级别 高中低	紧急情况/泼洒 特殊备注

过程步骤	化合物名称	类别	浓度	用量	化学品安全技术说明书	可能的伤害种类 或物理危害	降低或消除风险	技术	防护眼镜	手套	通风橱	清单/风险评估	可能性	危害	风险	危害级别	紧急情况/泼洒 特殊备注
D：铁含量的测定													1~4	1~4		高中低	
向上述预处理过的溶液中加水稀释至 150 mL,再加 15 mL H_2SO_4-H_3PO_4 混合酸溶液,随后加入 5~6 滴二苯胺磺酸钠指示剂,并立即用 $K_2Cr_2O_7$ 溶液滴定至溶液呈稳定的紫色即为终点。记录消耗的 $K_2Cr_2O_7$ 溶液体积,并做 3 组平行实验。	H_2SO_4-H_3PO_4 混合酸溶液	8	—	15 mL	√	皮肤灼伤和眼损伤	关注		√	√			1	2	2	低	大量水冲洗
实验产物																	
实验废弃物																	
整个过程的最终评估	低风险（按照正确流程操作）																
签名	实验老师：				实验室主任/职业安全健康（OHS）代表：							学生/操作人：					

Experiment 18　Determination of iron content in iron ores (mercury-free method)

【Objectives】

1. Master the basic principle and method of determining iron content by mercury-free method.

2. Learn the method of directly preparing $K_2Cr_2O_7$ standard solution, and its principle, method and experimental conditions for determining iron content, and further understand the application of $K_2Cr_2O_7$ in practical analysis.

3. Understand the method of acid decomposition of iron ore samples and the pretreatment process of REDOX, and clarify the importance of pretreatment.

4. Know the color-changing principle of REDOX indicators.

5. Enhance the awareness of environmental protection.

【Principles】

Iron ores include magnetite (Fe_3O_4), hematite (Fe_2O_3), siderite ($FeCO_3$), etc. $K_2Cr_2O_7$ analysis method has been generally adopted for determination of iron content in iron ores, which is simple and fast. However, considering the highly toxic of $HgCl_2$ that can be included in experimental wastewater, it is not good for reducing environmental pollution by using this traditional method. As a result, the improved potassium dichromate method was employed in this experiment without using $HgCl_2$.

1. Pretreatment of iron ores

Iron ore sample is firstly dissolved in hot concentrated HCl and $SnCl_2$ solution that is used to reduce most of the Fe(Ⅲ) to Fe(Ⅱ) while it is hot. The colour change of the test solution then happens from palm red to yellow. Next, take sodium tungstate as indicator, use $TiCl_3$ to reduce the residual Fe(Ⅲ). It will not end until the solution appears to be blue W(Ⅴ) (commonly known as tungsten blue). The corresponding reaction formula are as follows:

$$SnCl_4^{2-} + 2Fe^{3+} + 2Cl^- = SnCl_6^{2-} + 2Fe^{2+}$$

$$Fe^{3+} + Ti^{3+} + H_2O = TiO^{2+} + Fe^{2+} + 2H^+$$

The terminal point of the pretreatment process can be observed under the action of ene $K_2Cr_2O_7$ solution that will lead to blue disappearing.

2. Titration by $K_2Cr_2O_7$ solution

The pretreated solution is titrated with $K_2Cr_2O_7$ standard solution in sulfur-phosphorus mixed acid medium, and sodium diphenylamine sulfonate is used as indicator. The end point of titration does not reach until the solution becomes purple. The related reaction formula is:

$$Cr_2O_7^{2-} + 6Fe^{2+} + 14H^+ = 2Cr^{3+} + 6Fe^{3+} + 7H_2O$$

Thus, the content of Fe can be calculated as:

$$\omega = \frac{m_{Fe}}{G_{sample}} \times 100\% = \frac{6C_{Cr_2O_7^{2-}} \cdot V_{Cr_2O_7^{2-}} \times \frac{M_{Fe}}{1\ 000}}{G_{sample}} \times 100\%$$

Notably, the titration process above should be performed in the $H_2SO_4 - H_3PO_4$ mixed acid medium as:

(1) In order to make the electrode potential of ferroelectric (Fe^{3+}/Fe^{2+}) low enough to negative, and the titration jump range wider, so that the indicator's discoloration point falling within the jump range, the titration accuracy can be improved (sodium diphenylamine sulfonate is chosen as the indicator, and the titration end point is 0.85 V, falling outside 0.86~1.26).

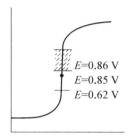

Figure 18-1 Diagram of titration curve

(2) To eliminate interference of Fe^{3+} to the end point observation.

【Apparatus and Chemicals】

1. Apparatus

Acid burette (50 mL); conical bottle (250 mL); volumetric bottle (250 mL); measuring cylinder (50 mL); beaker (500 mL); beaker (100 mL); watch glass; weighing bottle; dryer; analytical balance.

2. Chemicals

Iron ore solid; $K_2Cr_2O_7$ primary standard substance (being dried at 140 ℃ for 2 h); 10% $SnCl_2$ aqueous solution; $TiCl_3$ solution; HCl solution (6 mol·L^{-1}); $H_2SO_4 - H_3PO_4$ mixed acid solution; 25% Na_2WO_4 aqueous solution; 0.2% sodium diphenylamine sulfonate aqueous solution.

【Experimental Procedures】

1. Preparation of 0.017 mol·L^{-1} $K_2Cr_2O_7$ solution

Weigh 1.2~1.3 g $K_2Cr_2O_7$ reference reagent into a 100 mL beaker, add appropriate amount of water to dissolve, and then transfer to a 250 mL volumetric flask for constant volume.

2. Decomposition of iron ore sample

Accurately weighing 0.15~0.20 g iron ore sample and putting it in a 250 mL conical flask with a small amount of water wet, then adding 20 mL HCl solution and 8 to 10 drops of $SnCl_2$ solution (assistance of dissolving). For completely spreading the samples, the conical flask should be shaken 1~2 min before it is covered by watch glass. The resulted solution is then heated using nearly boiling water for 20~30 min to make the sample fully decomposed. After the residue becomes white, and the mixture is orange, washing the watch glass and the inner wall of the conical bottle with a little water.

3. Pretreatment of the iron ore sample

While the solution in the conical flask is hot, $SnCl_2$ solution is dripped to reduce Fe^{3+} until the solution changes from brown to light yellow. After that, 4 drops of Na_2WO_4 solution and 50 mL H_2O are added, followed by heating and shaking the flask. And $TiCl_3$ solution is added at the same time until the solution shows a stable light blue. Wash the the bottle wall with water to make it cool quickly to room temperature. Finally, add 2~5 drops of $K_2Cr_2O_7$ solution carefully to the obtained solution until the blue color just disappears.

4. Determination of iron content

The pretreated solution above is firstly diluted to 150 mL with water. Then 15 mL H_2SO_4-H_3PO_4 mixed acid solution is added, followed by adding 5~6 drops of sodium diphenylamine sulfonate indicator. After that, titration process is carried out immediately with $K_2Cr_2O_7$ solution until the solution shows a stable purple color, which is the end point. The consumed volume of $K_2Cr_2O_7$ solution is recorded and three parallel experiments are performed.

【Notes】

1. The decomposition of iron ore should be carried out at low temperature, because the boiling point of $FeCl_3$ is at 315 ℃ and it can volatilize significantly at about 100 ℃.

2. If the iron ore sample has been fully decomposed, SiO_2 should be the only one left and is white. Otherwise, it will be considered as incomplete decomposition.

3. To prevent Fe^{2+} from being oxidized, it is necessary to prepare one bottle of prereduced solution and one bottle of titrated solution. And the two bottles of solution cannot be prereduced at the same time.

4. During the quantitative reduction of Fe^{3+}, $SnCl_2$ or $TiCl_3$ cannot be used alone.

5. The volume of $SnCl_2$ solution should not be too large to avoid affecting the observation of Fe^{3+} color during the reduction of Fe^{3+}. In addition, the temperature should not be lower than 60 ℃, which can slow down the reaction and thus result in excessive $SnCl_2$ being not easy to remove.

6. During the titration process, the indicator of sodium diphenylamine sulfonate will consume a certain amount of $K_2Cr_2O_7$ solution, so it should not be added too much.

【Questions】

1. For quantitative reduction of Fe^{3+}, why does not use $SnCl_2$ or $TiCl_3$ alone? Why is it required to use the combined $SnCl_2$-$TiCl_3$ solution to prereduce? What are the disadvantages to use them alone?

2. What is the effect of adding phosphoric acid before titration? Why does titrate performed immediately after adding phosphoric acid?

Risk Assessment and Control Worksheet

TASK	Determination of iron content in iron ores (mercury-free method)				Assessor	Zeying WU, Hongsong WANG, Xingxing WU				Date	1 JUL. 2021			
Process Steps	Hazard Identification				Health Risk		Safety Measures				Checklist～Risk Assessment			Action
	Chemical Name	Class	Concn.	Amount	MSDS	Type of Injury Possible	To reduce or eliminate the risk	Technique	Glasses	Gloves	Fume Cbd.	Likely 1～4	Hazard 1～4	Risk Score	Risk Level H M L	In case of Emergency/Spill Special Comments

A: Preparation of 0.017 mol·L^{-1} K$_2$Cr$_2$O$_7$ solution

Chemical	Class	Concn.	Amount	MSDS	Injury	Reduce risk	Tech	Glasses	Gloves	Fume	Likely	Hazard	Score	Level	Action
K$_2$Cr$_2$O$_7$	9	0.017 M	250 mL	√	Skin burns and eye injuries	Handle with care		√	√		1	3	3	L	Sponge & water

B: Decomposition of iron ore sample

Chemical	Class	Concn.	Amount	MSDS	Injury	Reduce risk	Tech	Glasses	Gloves	Fume	Likely	Hazard	Score	Level	Action
HCl	2.3+8	6 M	20 mL	√	Skin burns and eye injuries	Handle with care		√	√	√	1	4	4	L	Sponge & water
SnCl$_2$	1	10%	1 mL	√	Skin burns and eye injuries	Handle with care		√	√	√	1	3	3	L	Sponge & water
Alcohol burner	—	—	—	—	Burns	Handle with care	√	√			1	3	3	L	Report immediately to Demonstrator or Prep Room

Accurately weighing 0.15～0.20 g iron ore sample and putting it in a 250 mL conical flask with a small amount of water wet, then adding 20 mL HCl solution and 8 to 10 drops of SnCl$_2$ solution (assistance of dissolving). For completely spreading the samples, the conical flask should be shaken 1～2 min before it is covered by watch glass. The resulted solution is then heated using nearly boiling water for 20～30 min to make the sample fully decomposed. After the residue becomes white, and the mixture is orange, washing the watch glass and the inner wall of the conical bottle with a little water.

continued

Risk Assessment and Control Worksheet

TASK	Assessor	Date
Determination of iron content in iron ores (mercury-free method)	Zeying WU, Hongsong WANG, Xingxing WU	1 JUL. 2021

Process Steps	Hazard Identification					Health Risk		Safety Measures				Checklist~Risk Assessment			Risk Assessment			Action	
	Chemical Name	Class	Concn.	Amount	MSDS	Type of Injury Possible	To reduce or eliminate the risk	Technique	Glasses	Gloves	Fume Cbd.	Likely	Hazard	Risk Score	Risk Level H M L	In case of Emergency/Spill	Special Comments
		or Physical Hazard											1~4	1~4				
	TiCl$_3$	—	15%~20%	—	√	Skin burns and eye injuries	Handle with care		√	√		1	3	3	L	Sponge & water	

C: Pretreatment of the iron ore sample

While the solution in the conical flask is hot, SnCl$_2$ solution is dripped to reduce Fe^{3+} until the solution changes from brown to light yellow. After that, 4 drops of Na$_2$WO$_4$ solution and 50 mL H$_2$O are added, followed by heating and shaking the flask. And TiCl$_3$ solution is added at the same time until the solution shows a stable light blue. Wash the the bottle wall with water to make it cool quickly to room temperature. Finally, add 2~5 drops of K$_2$Cr$_2$O$_7$ solution carefully to the obtained solution until the blue color just disappears.

continued

Risk Assessment and Control Worksheet

TASK	Determination of iron content in iron ores (mercury-free method)					Assessor	Zeying WU, Hongsong WANG, Xingxing WU				Date	1 JUL. 2021	
Process Steps	Hazard Identification					Health Risk	Safety Measures				Checklist～Risk Assessment		Action
	Chemical Name	Class	Concn.	Amount	MSDS	Type of Injury Possible	To reduce or eliminate the risk	Technique	Glasses	Gloves	Fume Cbd.	Risk Assessment	In case of Emergency/Spill
		or Physical Hazard										Likely / Hazard / Risk Score / Risk Level H M L	Special Comments
D: Determination of iron content												1～4 / 1～4	H M L
The pretreated solution above is firstly diluted to 150 mL with water. Then 15 mL H_2SO_4 – H_3PO_4 mixed acid solution is added, followed by adding 5～6 drops of sodium diphenylamine sulfonate indicator. After that, titration process is carried out immediately with $K_2Cr_2O_7$ solution until the solution shows a stable purple color, which is the end point. The consumed volume of $K_2Cr_2O_7$ solution is recorded and three parallel experiments are performed.	H_2SO_4 – H_3PO_4	8	—	15 mL	√	Skin burns and eye injuries	Handle with care		√	√		1 / 2 / 2	L / Sponge & water
Experimental Product (s)													
Experimental Waste (s)													
Final assessment of overall process	Low risk when correct procedures followed												
Signatures	Supervisor:					Lab Manager/OHS Representative:						Student/Operator:	

实验 19　$Na_2S_2O_3$ 标准溶液的配制与浓度标定及间接碘量法测定铜盐中的铜含量

【实验目的】

1. 掌握 $Na_2S_2O_3$ 标准溶液的配制方法。
2. 掌握 KIO_3 标定 $Na_2S_2O_3$ 溶液浓度的原理和方法,指示剂加入时间和溶液条件控制。
3. 熟悉碘量法测定的特点,了解酸度对反应和测定结果的影响。
4. 掌握碘量法测可溶性铜盐中铜含量的基本原理、操作条件和方法。

【实验原理】

1. 间接碘量法条件

间接碘量法的反应条件为中性或弱酸性,主要是由于:
(1) 强酸性介质中,$S_2O_3^{2-}$ 会分解析出单质硫。
(2) 碱性条件下,I_2 与 $S_2O_3^{2-}$ 反应除了生成 $S_4O_6^{2-}$ 外,还会生成 SO_4^{2-},同时 I_2 也会发生歧化反应。

2. $Na_2S_2O_3$ 溶液的配制

$Na_2S_2O_3$ 固体试剂常含有一些杂质,且易风化和潮解,溶液也不够稳定,容易分解,因此不能直接配制其标准溶液,而是要采用间接配制法。并且,配制好的 $Na_2S_2O_3$ 溶液需贮存于棕色瓶中,放置几天后再进行标定,长期使用的溶液应定期标定。

3. $Na_2S_2O_3$ 溶液的标定

标定 $Na_2S_2O_3$ 溶液的基准物有 $K_2Cr_2O_7$、KIO_3 和 $KBrO_3$ 等。
采用 KIO_3 为基准物,以淀粉溶液为指示剂,用间接碘量法(滴定碘法)标定 $Na_2S_2O_3$ 溶液。
(1) KIO_3 与过量的 KI 反应,析出与 KIO_3 计量相当的 I_2。

$$IO_3^- + 5I^- + 6H^+ = 3I_2 + 3H_2O$$

(2) 用 $Na_2S_2O_3$ 溶液滴定。

$$2S_2O_3^{2-} + I_2 = 2I^- + S_4O_6^{2-}$$

4. 铜盐中铜含量的测定——间接碘量法

弱酸性介质中,Cu^{2+} 与过量的 I^- 生成 CuI 沉淀,并定量地析出 I_2:

$$2Cu^{2+} + 4I^- = 2CuI\downarrow + I_2$$

为了使反应进行完全,I^- 必须加入过量,从而使得 I_2 转化成 I_3^-,增大 I_2 的溶解度:$2Cu^{2+} + 5I^- = 2CuI\downarrow + I_3^-$

反应生成的 I_2 以淀粉为指示剂,用 $Na_2S_2O_3$ 标准溶液滴定:

$$2S_2O_3^{2-} + I_2 \rightleftharpoons 2I^- + S_4O_6^{2-}$$

$$2S_2O_3^{2-} + I_3^- \rightleftharpoons 3I^- + S_4O_6^{2-}$$

由于 CuI 强烈地吸附 I_3^-,这部分 I_2 不与淀粉作用,从而使终点提前,使测定结果偏低,所以要在滴定终点前加入 SCN^-,使 CuI($K_{sp} = 1.1 \times 10^{-12}$)转化为溶解度更小的 CuSCN($K_{sp} = 4.8 \times 10^{-15}$),CuSCN 不能吸附 I_2,从而将吸附的 I_3^- 释放出来,提高了测定的准确度。

【器材与试剂】

1. 器材:250 mL 锥形瓶,500 mL 容量瓶,25.00 mL 移液管,50 mL 酸式滴定管,干燥器,分析天平。
2. 试剂:$K_2Cr_2O_7$ 基准试剂(在 140 ℃ 干燥 2 h),10% KI 溶液,$TiCl_3$ 溶液 0.10 mol·L^{-1} $Na_2S_2O_3$ 溶液,KSCN 溶液(100 g·L^{-1}),6 mol·L^{-1} HCl 溶液,1 mol·L^{-1} H_2SO_4 溶液,淀粉溶液(5 g·L^{-1},使用前配制),$CuSO_4 \cdot 5H_2O$,Na_2CO_3 试样。

【实验内容】

1. $Na_2S_2O_3$ 标准溶液的配制

称取 6.5 g $Na_2S_2O_3 \cdot 5H_2O$ 固体试样加水溶解,随后加入少许 Na_2CO_3 试样(约 0.5 g),溶解后加水稀释至 250 mL,将得到的溶液贮存于棕色玻璃瓶中,待标定。

2. $Na_2S_2O_3$ 标准溶液的标定

准确量取 20.00 mL KIO_3 标准溶液于 250 mL 锥形瓶中,分别加入 5 mL 6 mol·L^{-1} HCl 和 10 mL 10% KI,摇匀,盖上表面皿,于暗处放置 5 min 后再加 100 mL 水稀释并立即用 $Na_2S_2O_3$ 溶液滴定溶液呈浅黄绿色,随后加 2 mL 淀粉溶液指示剂并继续滴定至蓝色消失变为亮绿色即为终点。平行测定三次。

计算公式:$IO_3^- \hat{=} 3I_2 \hat{=} 6S_2O_3^{2-}$

$$C_{Na_2S_2O_3} = \frac{6(CV)_{KIO_3}}{V_{S_2O_3^{2-}}}$$

3. 铜盐中铜含量的测定

准确称取 0.5~0.6 g 铜盐三份于碘量瓶中,加 5 mL 1 mol·L^{-1} H_2SO_4 溶液和 100 mL 水使铜盐溶解,再加 10 mL 10% KI 溶液并立即用 $Na_2S_2O_3$ 溶液滴定至呈浅黄色,随后加入 2 mL 淀粉指示剂并继续滴定至溶液呈浅蓝色。加入 10 mL 100 g·L^{-1} 的 KSCN 溶液使得溶液蓝色转深(加入后,有 I_2 析出),继续用 KSCN 溶液滴定至蓝色刚好消失,此时即为滴定终点,且溶液呈米色或浅肉红色。平行测定三次。

计算公式:$2Cu^{2+} \hat{=} I_2 \hat{=} 2S_2O_3^{2-}$

$$\omega_{Cu} = \frac{(CV)_{S_2O_3^{2-}} \cdot M_{Cu}}{m_s} \times 100\%$$

【注意事项】

1. KIO_3 与 KI 的反应速率较慢,需要一定时间才能进行彻底,所以应先将溶液在暗处放置约 5 min,待反应完全后再进行滴定。

2. 酸度控制在 pH=3~4。酸度过低：会降低反应速度，同时 Cu^{2+} 部分水解（pH=8.0 时开始水解）。酸度过高：① I^- 易被空气中的氧氧化为 I_2（Cu^{2+} 有催化作用），使结果偏高；② $S_2O_3^{2-}$ 发生分解，$S_2O_3^{2-}+2H^+ \Longrightarrow S\downarrow +H_2SO_3$。

3. 淀粉指示剂应在临近终点时加入，而不能加入得过早。否则将有较多的 I_2 与淀粉指示剂结合，而这部分 I_2 在终点时解离较慢，造成终点拖后。

4. KSCN 溶液只能在临近终点时加入，否则大量 I_2 的存在有可能氧化 SCN^-，从而影响测定的准确度。

5. 实验要在碘量瓶中进行。

【思考题】

1. 如何配制和保存 I_2 溶液？配制 I_2 溶液时为什么要加入 KI？

2. 用 KIO_3 作基准物质标定 $Na_2S_2O_3$ 溶液时，为什么要加入过量的 KI 和 HCl 溶液？为什么要放置一段时间后才能加水稀释？为什么在滴定前还要加水稀释？

3. 若试样中含有铁，则加入何种试剂以消除铁对测定铜的干扰并控制溶液的 pH？

4. 碘量法的主要误差来源有哪些？为什么碘量法不适宜在高酸度或高碱度介质中进行？

- 视频演示
- 配套课件

风险评估及控制工作表

实验任务: $Na_2S_2O_3$ 标准溶液的配制与浓度标定及间接碘量法测定铜盐中的铜含量

评估人: 吴泽颖、王红松、吴星星

评估日期: 2021年7月1日

过程步骤	危害鉴别				评估健康风险		安全措施			清单/风险评估	风险评估			紧急情况/泼洒措施	特殊备注
	化合物名称	类别	浓度	用量	化学品安全技术说明书	可能的伤害种类	降低或消除风险	技术	防护眼镜	手套	通风橱		可能性 1~4	危害 1~4	风险	危害级别 高中低	

A: $Na_2S_2O_3$ 标准溶液的配制

称取 6.5 g $Na_2S_2O_3 \cdot 5H_2O$ 固体试样加水溶解,随后加入少许 Na_2CO_3 试样(约 0.5 g),将得到的溶液稀释至 250 mL,将得到的溶液贮存于棕色玻璃瓶中,待标定。

| | Na_2CO_3 | 2 | — | 0.5 g | √ | 眼睛和皮肤刺激 | 关注 | | | | | 1 | 3 | 3 | 低 | 大量水冲洗 | |

B: $Na_2S_2O_3$ 标准溶液的标定

准确量取 20.00 mL KIO_3 标准溶液于 250 mL 锥形瓶中,分别加入 5 mL、6mol·L^{-1} HCl 和 10 mL 10% KI,摇匀,盖上表面皿,于暗处放置 5 min 后再加 100 mL水稀释并立即用 $Na_2S_2O_3$ 溶液滴定溶液呈浅黄绿色,随后加 2 mL 淀粉指示剂继续滴定至蓝色消失变为亮绿色即为终点。平行测定三次。

	KI	—	10%	10 mL	√	—	关注		√	√	√	1	1	1	低	大量水冲洗	
	HCl	2,3+8	6 M	15 mL	√	皮肤灼伤和眼损伤	关注		√	√	√	1	4	3	低	大量水冲洗	
	KIO_3	5,1	0.1 M	20 mL	√	眼睛和皮肤刺激	关注		√	√	√	1	3	3	低	大量水冲洗	

续表

风险评估及控制工作表

实验任务	$Na_2S_2O_3$标准溶液的配制与浓度标定及间接碘量法测定铜盐中的铜含量				评估人	吴泽颖，王红松，吴星星				评估日期	2021年7月1日			
过程步骤	危害鉴别				健康风险	降低或消除风险	安全措施			清单/风险评估	风险评估			紧急情况/泼洒特殊备注
	化合物名称	类别	浓度	用量	化学品安全技术说明书	可能的伤害种类		技术	防护眼镜	手套	通风橱		可能性 1~4	危害 1~4	风险	危害级别 高中低

C：铜盐中铜含量的测定

准确称取0.5~0.6 g铜盐三份于碘量瓶中，加5 mL 1 mol·L^{-1} H_2SO_4溶液和100 mL水使铜盐溶解，再加10 mL 10% KI溶液并立即用$Na_2S_2O_3$溶液滴定至呈浅黄色，随后加入2 mL淀粉指示剂并继续滴定至溶液呈浅蓝色。加入10 mL 100 g·L^{-1}的KSCN溶液使得溶液蓝色转深（加入后，有I析出），继续滴KSCN溶液滴至蓝色刚好消失，此时即为滴定终点，且溶液呈米色或浅肉红色。平行测定三次。

化合物	类别	浓度	用量	化学品安全	可能的伤害	降低或消除	技术	防护眼镜	手套	通风橱	可能性	危害	风险	危害级别
H_2SO_4	8	1 M	15 mL	√	皮肤灼伤和眼损伤	关注		√	√	√	1	3	3	低
KSCN	—	100 g·L^{-1}	30 mL	√	皮肤接触及吞食有害	关注		√	√	√	1	4	4	低

紧急措施：大量水冲洗 / 大量水冲洗

实验产物：

实验废弃物：

整个过程的最终评估：低风险（按照正确流程操作）

签名： 实验老师： 实验室主任/职业安全健康（OHS）代表： 学生操作人：

Experiment 19 Preparation and calibration of $Na_2S_2O_3$ standard solution and determination of copper in copper content salt by indirect iodimetry method

【Objectives】

1. Master the preparation method of $Na_2S_2O_3$ standard solution.

2. Grasp the principle and method of calibrating the concentration of $Na_2S_2O_3$ solution by KIO_3, and learn to control the indicator addition time and solution conditions.

3. Be familiar with the characteristics of iodine assay and understand the influence of acidity on reaction and measurement results.

4. Master the basic principle, operating conditions and methods of measuring copper content in soluble copper salt by iodine measurement.

【Principles】

1. Conditions of indirect iodine method

The reaction condition of indirect iodimetry is neutral or slightly acidic, mainly due to:

(1) In strongly acidic medium, $S_2O_3^{2-}$ will decompose and form elemental sulfur.

(2) Under basic conditions, the reaction between I_2 and $S_2O_3^{2-}$ will generate not only $S_4O_6^{2-}$, but also SO_4^{2-}, and the disproportionation of I_2 will also occur.

2. Preparation of $Na_2S_2O_3$ solution

$Na_2S_2O_3$ sample contains some impurities, and is easy to be weathered and deliquesced. Besides, its corresponding solution is not stable enough and easy to decompose. As a result, $Na_2S_2O_3$ standard solution cannot be directly prepared, while indirect preparation method is required. Furthermore, the prepared $Na_2S_2O_3$ solution should be stored several days in the brown bottle before being calibrated. In addition, the solution for long-term use should be calibrated regularly.

3. Calibration of $Na_2S_2O_3$ solution

The reference materials for $Na_2S_2O_3$ solution include $K_2Cr_2O_7$, KIO_3 and $KBrO_3$ etc.

Herein, KIO_3 is adopted as the reference material and starch solution is used as the indicator. $Na_2S_2O_3$ solution will be calibrated by indirect iodimetry (titration iodimetry).

(1) KIO_3 is reacted with excessive KI, separating out I_2 that is measurement equivalent to $K_2Cr_2O_7$.

$$IO_3^- + 5I^- + 6H^+ = 3I_2 + 3H_2O$$

(2) Titration with $Na_2S_2O_3$ solution.

$$2S_2O_3^{2-} + I_2 = 2I^- + S_4O_6^{2-}$$

4. Determination of copper content in copper salt—indirect iodine content method

In weak acidic medium, Cu^{2+} is reacted with excess I^- to form CuI precipitation and quantitatively separate out I_2:

$$2Cu^{2+} + 4I^- = 2CuI\downarrow + I_2$$

In order to complete the reaction, excess I^- must be added to make I_2 transfer to I_3^-, and thus the solubility of I_2 can be increased: $2Cu^{2+} + 5I^- = 2CuI\downarrow + I_3^-$

The formed I_2 is titrated with $Na_2S_2O_3$ standard solution using starch as indicator:

$$2S_2O_3^{2-} + I_2 = 2I^- + S_4O_6^{2-}$$

$$2S_2O_3^{2-} + I_3^- = 3I^- + S_4O_6^{2-}$$

CuI will strongly adsorb I_3^-, thus, this part of I_2 does not act with starch, leading to the end point earlier and making the determination result lower. Therefore, SCN^- should be added before the end point of titration to convert CuI ($K_{sp} = 1.1 \times 10^{-12}$) into CuSCN ($K_{sp} = 4.8 \times 10^{-15}$) with less solubility. CuSCN cannot adsorb I_2, which can release the adsorptive I_3^- and improve the determination accuracy.

【Apparatus and Chemicals】

1. Apparatus

Conical bottle (250 mL); acid burette (50 mL); volumetric bottle (500 mL); pipette (25.00 mL); beaker; dryer; analytical balance.

2. Chemicals

KIO_3 primary standard substance (being dried at 140 ℃ for 2 h), 10% KI solution, $TiCl_3$ solution, 0.10 mol·L^{-1} $Na_2S_2O_3$ solution, KSCN solution (100 g·L^{-1}), HCl solution (6 mol·L^{-1}), H_2SO_4 solution (1 mol·L^{-1}), starch solution (5 g·L^{-1}, prepared before use), $CuSO_4 \cdot 5H_2O$ and Na_2CO_3 solid.

【Experimental Procedures】

1. Preparation of $Na_2S_2O_3$ solution

The solid sample $Na_2S_2O_3 \cdot 5H_2O$ of 6.5 g was accurately weighed and dissolved in water. Then a little Na_2CO_3 sample (about 0.5 g) was added, dissolved and the mixed solution is diluted with water to 250 mL. Finally, the resulted solution is stored in a brown glass bottle for calibration.

2. Calibration of $Na_2S_2O_3$ solution

Accurately measure 20 mL KIO_3 standard solution in a 250 mL conical flask, add 5 mL 6 mol·L^{-1} HCl solution and add 10 mL 10% KI solution, shake and cover the conical flask with watch glass, then place it in the dark for 5 min, add 100 mL H_2O to dilute the solution and then titrate it immediately with $Na_2S_2O_3$ solution until the solution appears light yellowish green, add 2 mL starch solution as indicator and continue titrate the solution until the blue disappears and it turns bright green. Perform three parallel measurements.

Formula: $IO_3^- \leftrightarrows 3I_2 \leftrightarrows 6S_2O_3^{2-}$

$$C_{Na_2S_2O_3} = \frac{6(CV)_{KIO_3}}{V_{S_2O_3^{2-}}}$$

3. Determination of copper content in the copper salt

Accurately weigh 0.5~0.6 g copper salt in the iodine flask (triplicate), and add 5 mL 1 mol·L^{-1} H_2SO_4 solution, 100 mL water to dissolve the copper salt, then add 10 mL 10% KI solution and titrate immediately with $Na_2S_2O_3$ solution until the test solution appears light yellow, add 2 mL starch solution as indicator and continue titrating the solution until it turns light blue, add 10 mL 100 g·L^{-1} KSCN solution. After the color of the test solution gets darker (I_2 precipitation), titrating with KSCN solution is continued until the blue color just disappears, which is the end of the titration, and the solution is beige or light meat red. Three parallel measurements should be carried out.

Formula: $2Cu^{2+} \leftrightarrows I_2 \leftrightarrows 2S_2O_3^{2-}$

$$\omega_{Cu} = \frac{(CV)_{S_2O_3^{2-}} \cdot M_{Cu}}{m_s} \times 100\%$$

[Notes]

1. The reaction rate of KIO_3 and KI is relatively slow, and it needs some time to be completed, thus, the solution should be placed in the dark for about 5 min before titration.

2. Keep acidity at pH=3~4.

(1) Too low acidity: it will slow down the reaction, while Cu^{2+} is partially hydrolyzed (pH=8.0).

(2) High acidity: I^- is easy to be oxidized by the oxygen in the air to I_2 (Cu^{2+} has catalytic effect), making the results higher. $S_2O_3^{2-}$ occurs decomposition: $S_2O_3^{2-} + 2H^+ \rightleftharpoons S\downarrow + H_2SO_3$.

3. The starch indicator should be added near the end. Otherwise, a lot of I_2 will bind to starch, which dissolves slowly at the end point and then causes the end point to be delayed.

4. KSCN solution can only be added near the end point. Otherwise, the presence of large amount of I_2 will lead to the oxidization of SCN^-, thus affecting the accuracy of determination.

5. The experiment should be carried out in an iodine measuring bottle.

[Questions]

1. How to prepare and preserve I_2 solution? Why is KI solution required for I_2 solution preparation?

2. Why should excessive KI and HCl solutions be added for calibration of $Na_2S_2O_3$ solution with KIO_3 as the reference material? Why should the solution be placed for a period of time before water dilution? Why is it necessary to be diluted before titration?

3. If there is a certain amount of iron contained in the sample, which reagent can be added to

eliminate the iron interference in the determination of copper content and controlling the pH value of the solution?

4. What are the main sources of error in iodimetry? Why is it not appropriate for iodimetry being operated in high acidity or high alkalinity media?

Risk Assessment and Control Worksheet

TASK	Assessor	Date
Preparation and concentration calibration of $Na_2S_2O_3$ standard solution and content determination of copper in copper salt by indirect iodimetry	Zeying WU, Hongsong WANG, Xingxing WU	1 JUL. 2021

Process Steps	Hazard Identification				Health Risk		Safety Measures				Checklist\Risk Assessment	Risk Assessment				Action	
	Chemical Name	Class	Concn.	Amount	MSDS	Type of Injury Possible	To reduce or eliminate the risk	Technique	Glasses	Gloves	Fume Cbd.		Likely	Hazard	Risk Score	Risk Level H M L	In case of Emergency/Spill	Special Comments
	or Physical Hazard												1~4	1~4				
A: Preparation of $Na_2S_2O_3$ solution																		
The solid sample $Na_2S_2O_3 \cdot 5H_2O$ of 6.5 g was accurately weighed and dissolved in water. Then a little Na_2CO_3 sample (about 0.5 g) was added, dissolved and the mixed solution is diluted with water to 250 mL. Finally, the resulted solution is stored in a brown glass bottle for calibration.	Na_2CO_3	2	—	0.5 g	√	Eye and skin irritant	Handle with care		√				1	3	3	L	Sponge & water
B: Calibration of Na_2CO_3 solution																		
Accurately measure 20 mL KIO_3 standard solution in a 250 mL conical flask, add 5 mL 6 mol·L^{-1} HCl solution and add 10 mL 10% KI solution, shake and cover the conical flask with watch glass, then place it in the dark for 5min, add 100 mL H_2O to dilute the solution and then titrate it immediately with $Na_2S_2O_3$ solution until the solution appears light yellowish green, add 2 mL starch solution as indicator and continue titrate the solution until the blue disappears and it turns bright green. Perform three parallel measurements.	KI	—	10%	10 mL	√	—	Handle with care		√	√			1	1	1	L	Sponge & water
	HCl	2.3+8	6 M	15 mL	√	Skin burns and eye injuries	Handle with care		√	√	√		1	3	3	L	Sponge & water
	KIO_3	5.1	0.1 M	20 mL	√	Eye and skin irrieant	Handle with care		√	√			1	3	3	L	Sponge & water

continued

<table>
<tr><td colspan="18">Risk Assessment and Control Worksheet</td></tr>
<tr><td>TASK</td><td colspan="5">Preparation and concentration calibration of $Na_2S_2O_3$ standard solution and content determination of copper in copper salt by indirect iodimetry</td><td>Assessor</td><td colspan="5">Zeying WU, Hongsong WANG, Xingxing WU</td><td>Date</td><td colspan="5">1 JUL. 2021</td></tr>
<tr><td>Process Steps</td><td colspan="5">Hazard Identification</td><td>Health Risk</td><td colspan="5">Safety Measures</td><td rowspan="3">Checklist～Risk Assessment</td><td colspan="4">Risk Assessment</td><td>Action</td></tr>
<tr><td rowspan="2"></td><td>Chemical Name</td><td>Class</td><td>Concn.</td><td>Amount</td><td>MSDS</td><td rowspan="2">Type of Injury Possible</td><td rowspan="2">To reduce or eliminate the risk</td><td rowspan="2">Technique</td><td rowspan="2">Glasses</td><td rowspan="2">Gloves</td><td rowspan="2">Fume Cbd.</td><td>Likely</td><td>Hazard</td><td rowspan="2">Risk Score</td><td>Risk Level</td><td>In case of Emergency/Spill</td></tr>
<tr><td colspan="5">or Physical Hazard</td><td>1～4</td><td>1～4</td><td>H M L</td><td>Special Comments</td></tr>
<tr><td colspan="18">C: Determination of copper content in the copper salt</td></tr>
<tr><td rowspan="2">Accurately weigh 0.5～0.6g copper salt in the iodine flask (triplicate), and add 5mL 1 mol・L^{-1} H_2SO_4 solution, 100ml water to dissolve the copper salt, then add 10 mL 10% KI solution and titrate immediately with $Na_2S_2O_3$ solution until the test solution appears light yellow, add 2 mL starch solution as indicator and continue titrating the solution until it turns light blue, add 10 mL 100 g・L^{-1} KSCN solution. After the color of the test solution gets darker (I_2 precipitation), titrating with KSCN solution is continued until the blue color just disappears, which is the end of the titration, and the solution is beige or light meat red. Three parallel measurements should be carried out.</td><td>H_2SO_4</td><td>8</td><td>1 M</td><td>15 mL</td><td>√</td><td>Skin burns and eye injuries</td><td>Handle with care</td><td></td><td>√</td><td>√</td><td>√</td><td></td><td>1</td><td>3</td><td>3</td><td>L</td><td>Sponge & water</td></tr>
<tr><td>KSCN</td><td>—</td><td>100 g・L^{-1}</td><td>30 mL</td><td>√</td><td>Harmful by skin contact</td><td>Handle with care</td><td></td><td>√</td><td>√</td><td>√</td><td></td><td>1</td><td>4</td><td>4</td><td>L</td><td>Sponge & water</td></tr>
</table>

continued

Risk Assessment and Control Worksheet										
TASK	Preparation and concentration calibration of Na₂S₂O₃ standard solution and content determination of copper in copper salt by indirect iodimetry	Assessor	Zeying WU, Hongsong WANG, Xingxing WU				Date	1 JUL. 2021		
Process Steps		Hazard Identification			Health Risk	Safety Measures		Risk Assessment		Action
		Chemical Name	Class or Physical Hazard	Concn.	Amount	MSDS	Type of Injury Possible	To reduce or eliminate the risk	Technique	Glasses	Gloves	Fume Cbd.	Checklist～Risk Assessment	Likely	Hazard	Risk Score	Risk Level	In case of Emergency/Spill	Special Comments
													1～4	1～4		H M L		
Experimental Product(s)										
Experimental Waste(s)										
Final assessment of overall process	Low risk when correct procedures followed									
Signatures	Supervisor:				Lab Manager/OHS Representative:				Student/Operator:	

实验 20　可溶性氯化物中氯含量的测定（莫尔法）

【实验目的】

1. 掌握 $AgNO_3$ 标准溶液的配制和标定。
2. 掌握莫尔法沉淀滴定的原理、方法和具体实验操作。

【实验原理】

某些可溶性氯化物中的氯含量常用莫尔法来测定。

莫尔法是在中性或者弱碱性溶液中，以 K_2CrO_4 为指示剂，用 $AgNO_3$ 标准溶液进行滴定。由于 AgCl 沉淀的溶解度比 Ag_2CrO_4 小，因此，AgCl 将首先从溶液中析出。当 AgCl 定量沉淀后，微过量的 $AgNO_3$ 溶液（1 滴）即与 CrO_4^{2-} 生成砖红色 Ag_2CrO_4 沉淀，它与白色的 AgCl 沉淀一起，使溶液略带橙红色，指示到达终点。

$$Ag^+ + Cl^- =\!=\!= AgCl\downarrow（白色）——滴定反应\ K_{sp}=1.8\times10^{-10}$$

$$2Ag^+ + CrO_4^{2-} =\!=\!= Ag_2CrO_4\downarrow（砖红色）——终点反应\ K_{sp}=2.0\times10^{-10}$$

【器材与试剂】

1. 器材：250 mL 锥形瓶，100 mL 容量瓶，25.00 mL 移液管，50 mL 酸式滴定管，干燥器，分析天平。
2. 试剂：K_2CrO_4 溶液（50 g·L^{-1}）、$AgNO_3$（分析纯）、NaCl（优级纯，使用前需在高温炉中 500～600 ℃ 下干燥 2～3 h 并贮存于干燥器内）。

【实验内容】

1. $AgNO_3$ 溶液的配制和标定

配制：称取 6.8 g $AgNO_3$ 固体样品，用少量水溶解后加水稀释至 400 mL 摇匀，将得到的溶液贮存于棕色试剂瓶中，待标定。

标定：准确称取 0.55～0.60 g NaCl 基准物质用水溶解后置于 100 mL 容量瓶中，稀释至刻度并充分摇匀。移取 25.00 mL 试样三份置于 250 mL 锥形瓶中，各加 20 mL 水和 1 mL 50 g·L^{-1} 的 K_2CrO_4 溶液，在不断摇动下用 $AgNO_3$ 溶液滴定至溶液呈微橙红色即为终点。计算 $AgNO_3$ 溶液的准确浓度。

2. 试样中 NaCl 含量的测定

准确称取 0.6～0.9 g 含氯试样，加水溶解后定量转移至 250 mL 容量瓶中，用水稀释至刻度并摇匀定容。移取 25.00 mL 试样三份置于 250 mL 锥形瓶中，各加 20 mL 水和 1 mL 50 g·L^{-1} 的 K_2CrO_4 溶液，在不断摇动下用 $AgNO_3$ 溶液滴定至溶液呈橙红色即为终点。根据试样的质量，$AgNO_3$ 溶液的浓度和滴定过程中消耗的体积,计算试样中 Cl^- 的含量。

计算公式：$$\omega_{Cl^-} = \frac{(CV)_{AgNO_3} \cdot M_{Cl}}{m_s \times \frac{25}{100}} \times 100\%$$

【注意事项】

1. 滴定应在中性或弱碱溶液中进行($pH = 6.5 \sim 10.5$)。

$$pH > 10.5: Ag^+ \longrightarrow AgOH \longrightarrow Ag_2O \downarrow$$

$$pH < 6.5: 2CrO_4^{2-} + 2H^+ \Longrightarrow 2HCrO_4^- \Longrightarrow Cr_2O_7^{2-} + H_2O$$

CrO_4^{2-} 转化为 $Cr_2O_7^{2-}$,降低$[Cr_2O_7^{2-}]$,使终点过迟。

2. $AgNO_3$ 需避光保存于棕色瓶中,并且不能与皮肤接触。
3. 滴定时要注意滴定速度并充分摇动以减少吸附。
4. 实验完毕后,为避免 AgCl 残留,需先用蒸馏水将装 $AgNO_3$ 溶液的滴定管冲洗 2~3 次,再用自来水洗净。

【思考题】

1. 配制好的 $AgNO_3$ 溶液要贮存于棕色瓶中,并放置于暗处,且使用过程中不能与肌肤接触,为什么?
2. 用 $K_2Cr_2O_7$ 作指示剂时,浓度过大或过小对测定有何影响?
3. 能否用莫尔法以 NaCl 标准溶液直接滴定 Ag^+? 为什么?

- 视频演示
- 配套课件

风险评估及控制工作表

实验任务	可溶性氯化物中氯含量的测定（莫尔法）				评估人	吴泽颖、王红松、吴星星			评估日期	2021 年 7 月 1 日			
过程步骤	危害鉴别				健康风险		安全措施			风险评估		紧急情况/泼洒措施	特殊备注
	化合物名称	类别	浓度	用量	化学品安全技术说明书	可能的伤害种类	降低或消除风险	技术	防护眼镜	手套	通风橱	清单/风险评估	可能性 1~4	危害 1~4	风险	危害级别 高中低	
A: $AgNO_3$ 溶液的配制和标定																	
配制：称取 6.8 g $AgNO_3$ 固体样品，用少量水溶解后加水稀释至 400 mL 摇匀，将得到的溶液贮存于棕色试剂瓶中，待标定。标定：准确称取 0.55~0.60 g NaCl 基准物质用水溶解后置于 100 mL 容量瓶中，稀释至刻度并充分摇匀。移取 25 mL 试液三份置于 250 mL 锥形瓶中，各加 20 mL 水和 1 mL 50 g·L^{-1} 的 K_2CrO_4 溶液，在不断摇动下用 $AgNO_3$ 溶液滴定至溶液呈橙红色即为终点，计算 $AgNO_3$ 溶液的准确浓度。	K_2CrO_4 溶液	—	50 g·L^{-1}	1 mL	√	皮肤/眼刺激	关注		√	√			1	2	2	低	大量水冲洗
	$AgNO_3$	5.1	—	6.8 g	√	皮肤烧伤和眼损伤	关注		√	√			1	3	3	低	大量水冲洗
B: 试样中 NaCl 含量的测定																	
准确称取 0.6~0.9 g 含氯试样，加水溶解后定量转移至 250 mL 容量瓶中，用水稀释至刻度并定容。移取 25 mL 试样三份置于 250 mL 锥形瓶中，各加 20 mL 水和 1 mL 50 g·L^{-1} 的 K_2CrO_4 溶液，在不断摇动下用 $AgNO_3$ 溶液滴定至溶液呈橙红色即为终点。根据试样的质量、$AgNO_3$ 溶液的浓度和滴定过程中消耗的体积，计算试样中 Cl^- 的含量。	$AgNO_3$	5.1	—		√	皮肤灼伤和眼损伤	关注		√	√			1	3	3	低	大量水冲洗

续 表

风险评估及控制工作表

实验任务	可溶性氯化物中氯含量的测定（莫尔法）			评估人	吴泽颖、王红松、吴星星			评估日期	2021年7月1日			
过程步骤	危害鉴别			健康风险	安全措施			清单/风险评估	风险评估			措施
	化合物名称	类别	浓度	用量	化学品安全技术说明书	可能的伤害种类或物理危害	降低或消除风险	技术	防护眼镜	手套	通风橱		可能性	危害	风险	危害级别	紧急情况/泼洒	特殊备注
													1~4	1~4		高中低		
实验产物												
实验废弃物												
整个过程的最终评估	低风险（按照正确流程操作）											
签名	实验老师：			实验室主任/职业安全健康（OHS）代表：				学生/操作人：				

234

Experiment 20　Determination of chlorine content in soluble chlorides (Mohr method)

【Objectives】

1. To master the preparation and calibration of $AgNO_3$ standard solution.
2. To grasp the principle, method and specific experimental operation of Mohr method.

【Principles】

Mohr method is generally adopted for determination of chlorine content in those soluble chlorides.

Mohr method is performed in neutral or weakly alkaline solutions, using K_2CrO_4 as indicator. And the titration process is under the action of $AgNO_3$ standard solution. Since the solubility of AgCl precipitation is weaker than that of Ag_2CrO_4, AgCl will first be precipitated from the solution. After AgCl is quantitatively precipitated, the slightly excessive $AgNO_3$ solution (about 1 drop) is precipitated with CrO_4^{2-} into brick-red Ag_2CrO_4, which, together with the white AgCl precipitate, gives the solution a reddish orange color, indicating the titration destination.

$$Ag^+ + Cl^- =\!=\!= AgCl \downarrow \text{(White)} \quad\text{——}\quad \text{Titration reaction } K_{sp}=1.8\times10^{-10}$$

$$2Ag^+ + CrO_4^{2-} =\!=\!= Ag_2CrO_4 \downarrow \text{(Brick-red)} \quad\text{——}\quad \text{End reaction } K_{sp}=2.0\times10^{-10}$$

【Apparatus and Chemicals】

1. Apparatus

Conical bottle (250 mL); acid burette (50 mL); volumetric bottle (100 mL); pipette (25.00 mL); beaker; dryer; analytical balance.

2. Chemicals

K_2CrO_4 solution (50 g·L^{-1}), $AgNO_3$(AR), NaCl (GR, being dried at 500~600 ℃ for 2~3 and stored in the dryer before use).

【Experimental Procedures】

1. Preparation and calibration of $AgNO_3$ solution

Preparation: About 6.8 g $AgNO_3$ solid is weighed and dissolved with a small amount of water before being diluted to 400 mL. After shaken well, the resulting solution was stored in a brown reagent bottle for calibration.

Calibration: 0.55~0.60 g NaCl reference material is accurately weighed and dissolved in water, and then transferred into a 100 mL volumetric flask, followed by being diluted to the scale and shaken well. 25.00 mL sample (three copies) is taken and placed in a 250 mL conical flask. Subsequently, 20 mL water and 1 mL 50 g·L^{-1} K$_2$CrO$_4$ solution are added. With constant shaking, AgNO$_3$ solution is used to titrate the solution until it is reddish orange, which is the end point. The consumed volume of AgNO$_3$ solution should be recored.

2. Determination of NaCl content

0.6~0.9 g chlorine-containing sample is accurately weighed, dissolved in water and then quantitatively transferred to a 250 mL volumetric flask. The solution is diluted with water to the scale and shaken to constant volume. After that, 25.00 mL sample is transferred into a 250 mL conical flask, with 20 mL water and 1 mL 50 g·L^{-1} K$_2$CrO$_4$ solution added. AgNO$_3$ solution is used to titrate the solution with constant shaking until it turns orange red. According to the mass of the sample, the concentration of AgNO$_3$ solution and the volume consumed during the titration, the content of Cl$^-$ in the chloride sample can be calculated. The measurement will be operated four times in parallel.

The computational formula: $\omega_{Cl^-} = \dfrac{(CV)_{AgNO_3} \cdot M_{Cl}}{m_s \times \dfrac{25}{100}} \times 100\%$

【Notes】

1. Titration should be performed in neutral or weak base solution (pH=6.5~10.5).

$$pH > 10.5: Ag^+ \longrightarrow AgOH \longrightarrow Ag_2O \downarrow$$

$$pH < 6.5: 2CrO_4^{2-} + 2H^+ \rightleftharpoons 2HCrO_4^- \rightleftharpoons Cr_2O_7^{2-} + H_2O$$

CrO$_4^{2-}$ is converted to Cr$_2$O$_7^{2-}$, then the concentration of CrO$_4^{2-}$ decreases, making the end point too late.

2. AgNO$_3$ must be kept hidden from light, stored in a brown bottle, and not in contact with skin.

3. Pay attention to the titration speed and shake enough to reduce adsorption.

4. After the experiment, the burette containing AgNO$_3$ solution needs to be rinsed 2~3 times with distilled water and then cleaned with tap water to avoid AgCl residue.

【Questions】

1. Why should the prepared AgNO$_3$ solution be stored in the brown bottle, kept in the dark and not in contact with the skin during use?

2. What is the effect if the concentration of K$_2$Cr$_2$O$_7$ indicator is too high or too low?

3. Can Ag$^+$ be directly used to titrate NaCl standard solution by Mohr method? Why?

Risk Assessment and Control Worksheet

TASK	Determination of chlorine content in soluble chlorides (Mohr method)					Assessor	Zeying WU, Hongsong WANG, Xingxing WU				Date	1 JUL. 2021	
Process Steps	Hazard Identification				Health Risk	Safety Measures				Checklist╲Risk Assessment	Risk Assessment		Action
	Chemical Name	Class	Concn.	Amount	MSDS	Type of Injury Possible	To reduce or eliminate the risk	Technique	Glasses	Gloves	Fume Cbd.	Likely	Hazard	Risk Score	Risk Level H M L	In case of Emergency/Spill Special Comments

Process Steps	Chemical Name	Class	Concn.	Amount	MSDS	Type of Injury Possible	To reduce or eliminate the risk	Technique	Glasses	Gloves	Fume Cbd.	Likely 1~4	Hazard 1~4	Risk Score	Risk Level H M L	In case of Emergency/Spill Special Comments
A: Preparation and calibration of AgNO$_3$ solution																
Preparation: About 6.8 g AgNO$_3$ solid is weighed and dissolved with a small amount of water before being diluted to 400 mL. After shaken well, the resulting solution was stored in a brown reagent bottle for calibration. Calibration: 0.55~0.60 g NaCl reference material is accurately weighed and dissolved in water, and then transferred into a 100 mL volumetric flask, followed by being diluted to the scale and shaken well. 25 mL sample (three copies) is taken and placed in a 250 mL conical flask. Subsequently, 20 mL water and 1 mL 50 g · L^{-1} K$_2$CrO$_4$ solution are added. With constant shaking, AgNO$_3$ solution is used to titrate the solution until it is reddish orange, which is the end point. The consumed volume of AgNO$_3$ solution should be recored.	K$_2$CrO$_4$	—	50 g · L^{-1}	1 mL	✓	Skin and eye irritant	Handle with care		✓	✓		1	2	2	L	Sponge & water
	AgNO$_3$	5.1	—	6.8 g	✓	Skin burns and eye injuries	Handle with care		✓	✓		1	3	3	L	Sponge & water
B: Determination of NaCl content																

Risk Assessment and Control Worksheet

TASK	Assessor	Date
Determination of chlorine content in soluble chlorides (Mohr method)	Zeying WU, Hongsong WANG, Xingxing WU	1 JUL. 2021

Process Steps	Hazard Identification or Physical Hazard					Health Risk		Safety Measures				Checklist～Risk Assessment	Risk Assessment			Action	
	Chemical Name	Class	Concn.	Amount	MSDS	Type of Injury Possible	To reduce or eliminate the risk	Technique	Glasses	Gloves	Fume Cbd.		Likely 1～4	Hazard 1～4	Risk Score	Risk Level H M L	In case of Emergency/Spill	Special Comments
0.6～0.9 g chlorine-containing sample is accurately weighed, dissolved in water and then quantitatively transferred to a 250 mL volumetric flask. The solution is diluted with water to the scale and shaken to constant volume. After that, 25 mL sample is transferred into a 250 mL conical flask, with 20 mL water and 1 mL 50 g·L^{-1} K$_2$CrO$_4$ solution added. AgNO$_3$ solution is used to titrate the solution with constant shaking until it turns orange red. According to the mass of the sample, the concentration of AgNO$_3$ solution and the volume consumed during the titration, the content of Cl$^-$ in the chloride sample can be calculated. The measurement will be operated four times in parallel.	AgNO$_3$	5.1	—	—	✓	Skin burns and eye injuries	Handle with care		✓	✓			1	3	3	L	Sponge & water

Experimental Product (s)	
Experimental Waste (s)	
Final assessment of overall process	Low risk when correct procedures followed

Signatures	Supervisor:	Lab Manager/OHS Representative:	Student/Operator:

第 6 章
综合性、设计性、研究性实验

实验 21　硫酸亚铁铵的制备

【实验目的】

1. 初步了解无机固体化合物的制备方法。
2. 掌握倾析法的原理。
3. 了解硫酸亚铁铵计量比。
4. 初步掌握结晶法的应用。

【实验原理】

使用铁屑为原料,与稀硫酸反应生成硫酸亚铁。后经过硫酸铵调节 pH,在溶液中按计量比生成硫酸盐亚铁铵。经浓缩,可析出硫酸亚铁铵晶体。

反应的化学方程式:

$$Fe + H_2SO_4 = FeSO_4 + H_2\uparrow$$

$$FeSO_4 + (NH_4)_2SO_4 + 6H_2O = FeSO_4 \cdot (NH_4)_2SO_4 \cdot 6H_2O$$

【器材与试剂】

1. 器材:烧杯,玻璃棒,天平,滴管,水浴锅,蒸发皿,酒精灯,三脚架等。
2. 试剂:铁屑,稀硫酸,硫酸铵,碳酸钠,95%乙醇,广泛 pH 试纸。

【实验内容】

称取铁屑 2.0 g,置于烧杯中。加 20 mL 碳酸钠溶液并加热搅拌 10 min,以除去铁屑表面的油污。用倾析法倒出上层清液,后加蒸馏水 30 mL,洗涤铁屑。用倾析法倒出上层清液。重复上述步骤三次,直至洗净铁屑。量取硫酸溶液 10 mL，加入铁屑中，再加 10 mL 蒸馏水。置于水浴锅中加热，温度控制在 80 ℃，反应至无气泡产生。过滤得澄清透明的淡绿色溶液为 $FeSO_4$ 溶液。使用硫酸铵调节 pH 至 3。将溶液至于三脚架上，加热蒸发，至溶液有少量晶膜产生。停止加热，冷却至室温，让硫酸亚铁铵缓慢从蒸发皿中析出。后过滤晶体，使用 95% 乙醇洗涤，得到浅绿色晶体，为硫酸亚铁铵。称重并计算产率。

【注意事项】

1. 使用硫酸时务必小心。
2. 加热蒸发时需要不断搅拌，防止溶液飞溅。
3. 静置结晶时，应缓慢冷却。

【思考题】

1. 硫酸铵过量会有什么结果？
2. 为什么铁屑不需要完全溶解于硫酸中？

● 视频演示

风险评估及控制工作表

实验任务	硫酸亚铁铵的制备				评估人	吴泽颖,王红松,吴昊星星				评估日期	2021年7月1日			
过程步骤	危害鉴别				健康风险	降低或消除风险	安全措施			清单/风险评估	风险评估			紧急情况、泼洒、特殊备注
	化合物名称	类别	液度	用量	化学品安全技术说明书	可能的伤害种类		技术	防护眼镜	手套	通风橱		可能性 危害 风险	危害级别 高中低

过程步骤	化合物名称	类别	液度	用量	化学品安全技术说明书	可能的伤害种类	降低或消除风险	技术	防护眼镜	手套	通风橱	可能性 1~4	危害 1~4	风险	危害级别	紧急情况、泼洒、特殊备注
A:称铁屑 2.0 g,置于烧杯中。加 20 mL 碳酸钠溶液并加热搅拌 10 min,以除去铁屑表面的油污。	碳酸钠	—	—	2.0 g	√	眼刺激	关注		√	√	√	1	3	3	低	大量水冲洗
B:量取硫酸溶液 10 mL,再加 10 mL 蒸馏水。置于铁屑中,加入铁锅中加热,温度控制在 80 ℃,反应至无气泡产生。	硫酸	8	3 M	10 mL	√	皮肤灼伤和眼损伤	关注		√	√	√	1	3	3	低	大量水冲洗
C:过滤得澄清透明的淡绿色溶液至 FeSO₄ 溶液。使用硫酸铵调节 pH 至 3。	硫酸铵	—	—	滴加	√	—	关注		√	√	√	1	2	2	低	大量水冲洗
上:加热蒸发,至溶液有少量晶膜产生。	加热装置	—	—	—		电击&烫伤	确保电线贴上标签	√	√	√		1	3	3	低	立即报告实验老师或实验备员
D:过滤晶体,使用 95%乙醇洗涤,得到浅绿色晶体,为硫酸亚铁铵。称重并计算产率。	乙醇	3	95%	10 mL	√	—	关注		√	√	√	1	2	2	低	大量水冲洗
实验产物	硫酸亚铁铵	—	—	1.8 g	√	皮肤刺激	关注		√	√		1	1	1	低	用大量水冲洗
实验废弃物																

整个过程的最终评估: 低风险(按照正确流程操作)

签名: 实验老师: 实验室主任/职业安全健康(OHS)代表: 学生/操作人:

Experiment 21　Preparation of ammonium iron(Ⅱ) sulfate

【Objectives】

1. Preliminary understanding the preparation method of inorganic solid compounds.
2. Master the principles of decantation.
3. Understand the metering ratio of ammonium ferrous sulfate.
4. Master the application of crystallization method.

【Principles】

Using iron filings as raw materials, it reacts with dilute sulfuric acid to produce iron sulfite. After that, the pH value is adjusted by ammonium sulfate, and ammonium ferrous sulfate is generated in the solution at a metering ratio. After concentration, crystals of ferrous ammonium sulfate can be precipitated.

Reaction equation:

$$Fe + H_2SO_4 = FeSO_4 + H_2 \uparrow$$

$$FeSO_4 + (NH_4)_2SO_4 + 6H_2O = FeSO_4 \cdot (NH_4)_2SO_4 \cdot 6H_2O$$

【Apparatus and Chemicals】

1. Apparatus

Beaker; glass rod; balance; dropper; water bath; evaporation dish; alcohol lamp; tripod; etc.

2. Chemicals

Iron filings; dilute sulfuric acid; ammonium sulfate; sodium carbonate; 95% ethanol; extensive pH test paper.

【Experimental Procedures】

Weigh 2.0 g of iron filings and place in a beaker. Add 20 mL of sodium carbonate solution and heat and stir for 10 min to remove the oil on the surface of iron filings. The supernatant was decanted by decantation, and 30 mL of distilled water was added to wash iron filings. The supernatant was decanted. Repeat the above steps three times until the iron filings are washed. Measure 10 mL of sulfuric acid solution, add to iron filings, and add 10 mL of distilled water. Put it in a water bath and heat it. The temperature is controlled at 80 ℃. Filtration gave a clear and transparent light green solution as $FeSO_4$ solution. Adjust the pH to 3 using ammonium sulfate. Place the solution on a tripod, heat and evaporate until a small amount of crystal film is generated in the solution. Stop heating and cool to room temperature to allow ammonium ferrous sulfate to slowly precipitate out of the evaporation dish. The crystals were then filtered and washed with 95% ethanol to obtain pale green crystals as ammonium ferrous sulfate. Weigh and calculate the yield.

【Notes】

1. Using sulfuric acid carefully.
2. Constant stirring is required during heating and evaporation to prevent the solution from splashing.
3. Cooling slowly to crystallize.

【Questions】

1. What is the result of excessive ammonium sulfate?
2. Why doesn't iron filings need to be completely dissolved in sulfuric acid?

Risk Assessment and Control Worksheet

TASK	Preparation of Ammonium iron(Ⅱ) sulfate				Assessor	Zeying WU, Hongsong WANG, Xingxing WU					Date	1 JUL. 2021			
Process Steps	Hazard Identification				Health Risk		Safety Measures				Risk Assessment			Action	
	Chemical Name	Class	Concn.	Amount	MSDS	Type of Injury Possible	To reduce or eliminate the risk	Technique	Glasses	Gloves	Fume Cbd.	Checklist＼Risk Assessment	Likely 1~4	Hazard 1~4	Risk Score	Risk Level H M L	In case of Emergency/Spill	Special Comments

Process Steps	Chemical Name	Class	Concn.	Amount	MSDS	Type of Injury Possible	To reduce or eliminate the risk	Technique	Glasses	Gloves	Fume Cbd.	Likely	Hazard	Risk Score	Risk Level	In case of Emergency/Spill
			or Physical Hazard													
A: Weigh 2.0 g of iron filings and place in a beaker. Add 20 mL of sodium carbonate solution and heat and stir for 10 min to remove the oil on the surface of iron filings.	Sodium carbonate	—	—	2.0 g	✓	Eye irritant	Handle with care		✓	✓		1	3	3	L	Sponge & water
B: Measure 10 mL of sulfuric acid solution, add to iron filings, and add 10 mL of distilled water. Put it in a water bath and heat it. The temperature is controlled at 80 ℃.	Sulfuric acid	8	3 M	10 mL	✓	Skin burns and eye injuries	Handle with care		✓	✓	✓	1	3	3	L	Sponge & water
C: Adjust the pH to 3 using ammonium sulfate. Place the solution on a tripod. Heat and evaporate until a small amount of crystal film is generated in the solution.	Ammonium sulphate	—	—	2 drops	✓	—	Handle with care		✓	✓	✓	1	2	2	L	Sponge & water
	Heating device					Electric shock & burns	Handle with care	✓	✓	✓	✓	1	3	3	L	Report immediately to Demonstrator or Prep Room
D: Washed with 95% ethanol to obtain pale green crystals as ammonium ferrous sulfate.	Ethanol	3	95%	10 mL	✓	—	Handle with care		✓	✓	✓	1	2	2	L	Sponge & water
Experimental Product (s)	Ferrous ammonium sulfate	—		1.8 g	✓	Skin irritant	Handle with care		✓	✓		1	1	1	L	Sponge & water

Experimental Waste (s)

Final assessment of overall process: Low risk when correct procedures followed

Signatures Supervisor: Lab Manager/OHS Representative: Student/Operator:

实验 22　三草酸合铁酸钾的制备

【实验目的】

1. 了解无机固体合成方法。
2. 熟悉 Fe 价态变化。
3. 掌握结晶法的原理和规律。

【实验原理】

使用硫酸亚铁铵为原料,与草酸反应生成草酸亚铁。通过过氧化氢氧化,使 Fe^{2+} 转化为 Fe^{3+}。经过草酸合草酸钾调节 pH,得到翠绿色的三草酸合铁酸钾的溶液。后使用乙醇和冷却结晶技术将三草酸合铁酸钾从溶液中析出。

$$FeSO_4 \cdot (NH_4)_2SO_4 \cdot 6H_2O + H_2C_2O_4 \longrightarrow FeC_2O_4 \cdot 2H_2O \downarrow + (NH_4)_2SO_4$$
$$2FeC_2O_4 \cdot 2H_2O + H_2O_2 + 3K_2C_2O_4 + H_2C_2O_4 \longrightarrow 2K_3[Fe(C_2O_2)]_3 \cdot 3H_2O$$

【器材与试剂】

1. 器材:烧杯,玻璃棒,天平,滴管,水浴锅,蒸发皿,酒精灯,三脚架等。
2. 试剂:硫酸亚铁铵,过氧化氢,草酸钾,草酸,乙醇,广泛 pH 试纸。

【实验内容】

称取 4.0 g $FeSO_4 \cdot 7H_2O$ 晶体,放入烧杯中再加入 15 mL 去离子水,加热溶解。再加入 1 mol·L^{-1} $H_2C_2O_4$ 20 mL,搅拌并加热煮沸,使形成淡黄色沉淀 FeC_2O_4,用倾析法洗涤该沉淀 3 次,每次使用 25 mL H_2O 去除可溶性杂质。

在上述沉淀中加入 10 mL 饱和 $K_2C_2O_4$ 溶液,水浴加热至 40 ℃,滴加 3% H_2O_2 溶液 20 mL,不断搅拌溶液并维持温度在 40℃左右,使 Fe(Ⅱ)充分氧化为 Fe(Ⅲ)。滴加完后,加热溶液至沸腾以去除过量的 H_2O_2。

保持上述沉淀近沸状态,先加入 1 mol·L^{-1} $H_2C_2O_4$ 7 mL,然后趁热滴加 1 mol·L^{-1} $H_2C_2O_4$ 1~2 mL 使沉淀溶解,溶液的 pH 保持在 4~5,此时溶液呈翠绿色,趁热将溶液过滤到一个 150 mL 烧杯中,并使滤液控制在 30 mL 左右,冷却放置过夜、结晶、抽滤至干即得三草酸合铁(Ⅲ)酸钾晶体。称量,计算产率,并将晶体置于干燥器内避光保存。

【思考题】

1. 得到翠绿色溶液后,冰水浴和普通室温冷却现象有什么不同?
2. 过氧化氢若不加热除去,会有什么影响?

风险评估及控制工作表

实验任务：三草酸合铁酸钾的制备
评估人：吴泽颖、王红松、吴星星
评估日期：2021年7月1日

过程步骤	危害鉴别					评估风险		安全措施				风险评估			紧急情况/泼洒特殊备注
	化合物名称	类别	浓度	用量	化学品安全技术说明书	可能的伤害种类	降低或消除风险	技术	防护眼镜	手套	通风橱	清单/风险评估	可能性 1~4	危害 1~4	风险	危害级别
A：I）称取 4.0 g FeSO$_4$·7H$_2$O 晶体，放入烧杯中再加入15 mL去离子水。	硫酸亚铁	—		4.0 g	√	皮肤/眼刺激	关注		√	√			1	1	1	高中低
II）加热溶解。	加热装置	戒物理危害				烫伤 & 电击	确保电线已贴好标签						1	3	3	低
B：加入 1 mol·L^{-1} H$_2$C$_2$O$_4$ 20 mL，搅拌并加热煮沸，使形成淡黄色沉淀FeC$_2$O$_4$，用倾析法洗涤该沉淀3次，每次使用25 mL H$_2$O去除可溶性杂质。	草酸	—	1 M	20 mL	√	吞咽/皮肤接触有害	关注		√	√	√		1	2	2	低
C：I）上述沉淀中加入 10 mL 饱和 K$_2$C$_2$O$_4$ 溶液，水浴加热至40℃。	草酸钾	—	饱和溶液	10 mL		吞咽有害	关注		√	√	√		1	2	2	低
II）滴加 3% H$_2$O$_2$ 溶液 20 mL，不断搅拌溶液并维持温度在40℃左右，使 Fe(II)充分氧化为 Fe(III)。	过氧化氢	5.1+8	3%	20 mL	√	皮肤灼伤和眼损伤	关注		√	√	√		1	3	3	低
III）滴加完毕后，加热溶液至沸腾以去除过量的 H$_2$O$_2$。	加热装置	—				烫伤 & 电击	确保电线已贴好标签						1	3	3	低

应急措施：大量水冲洗；立即向实验老师或准备人员报告

续 表

风险评估及控制工作表

实验任务	三草酸合铁酸钾的制备				评估人	吴泽颖,王红松,吴星星				评估日期	2021 年 7 月 1 日			
过程步骤	危害鉴别				健康风险	安全措施				清单/风险评估	风险评估			紧急情况/泼洒特殊备注
	化合物名称	类别或物理危害	浓度	用量	化学品安全技术说明书	可能的伤害种类	降低或消除风险	技术	防护眼镜	手套	通风橱	可能性 1~4	危害 1~4	风险	危害级别 高中低	措施
D: 先加入 1 mol·L⁻¹ H₂C₂O₄ 7 mL,然后趁热滴加 1 mol·L⁻¹ H₂C₂O₄ 1~2 mL,使沉淀溶解,溶液的 pH 保持在 4~5,此时溶液呈翠绿色,趁热将溶液过滤到一个 150 mL 烧杯中,并使滤液控制在 30 mL 左右。	草酸	—	1 M	20 mL	√	吞咽/皮肤接触有害	关注		√	√		1	2	2	低	大量水冲洗
实验产物	三草酸合铁酸钾	—	—	2.0 g	√	—	关注		√	√		1	1	1	L	用大量水冲洗
实验废弃物																
整个过程的最终评估	低风险(按照正确流程操作)															
签名	实验老师:				实验室主任/职业安全健康(OHS)代表:							学生/操作人:				

Experiment 22   Preparation of tripotassium trioxalatoferrate

【Objectives】

1. Understand inorganic solid synthesis methods.
2. Familiar with Fe valence changes.
3. Master principles and laws of crystallization.

【Principles】

Using ferrous ammonium sulfate as raw material, it reacts with oxalic acid to form ferrous oxalate. By hydrogen peroxide oxidation, Fe^{2+} is converted to Fe^{3+}. After adjusting the pH value by potassium oxalate oxalate, a solution of emerald green potassium ferrate oxalate was obtained. Triethanol and potassium ferrite were then precipitated from the solution using ethanol and cooling crystallization techniques.

$$FeSO_4 \cdot (NH_4)_2SO_4 \cdot 6H_2O + H_2C_2O_4 \longrightarrow FeC_2O_4 \cdot 2H_2O \downarrow + (NH_4)_2SO_4$$

$$2FeC_2O_4 \cdot 2H_2O + H_2O_2 + 3K_2C_2O_4 + H_2C_2O_4 \longrightarrow 2K_3[Fe(C_2O_2)]_3 \cdot 3H_2O$$

【Apparatus and Chemicals】

1. Apparatus

Beaker, glass rod, balance, dropper, water bath, evaporation dish, alcohol lamp, tripod, etc.

2. Chemicals

Ammonium ferrous sulfate, hydrogen peroxide, potassium oxalate, oxalic acid, ethanol, wide pH test paper.

【Experimental Procedures】

Weigh 4.0 g of $FeSO_4 \cdot 7H_2O$ crystals, put them into a beaker and add 15 mL of deionized water, heat to dissolve. Add 1 mol·L^{-1} $H_2C_2O_4$ 20 mL, stir and heat to boil to form a light yellow precipitate FeC_2O_4, wash the precipitate 3 times by decantation, and use 25 mL H_2O each time to wash away soluble impurities.

10 mL of saturated $K_2C_2O_4$ solution was added to the above precipitate, heated to 40 ℃ in a water bath, and 20 mL of a 3% H_2O_2 solution was added dropwise. The solution was continuously stirred

and maintained at a temperature of about 40 ℃ to fully oxidize Fe(Ⅱ) to Fe(Ⅲ). After the addition was complete, the solution was heated to boiling to remove excess H_2O_2.

To maintain the near-boiling state of the precipitate, first add 1 mol·L^{-1} $H_2C_2O_4$ 7 mL, and then dropwise add 1 mol·L^{-1} $H_2C_2O_4$ 1~2 mL while hot to dissolve the precipitate. The pH value of the green solution is maintained at 4~5, filter the solution into a 150 mL beaker while it is still hot, and control the filtrate to about 30 mL. Cool to room temperature, crystallize, and filter to dryness to obtain potassium iron (Ⅲ) trioxalate crystals. Weigh, calculate the yield, and store the crystals in a desiccator protected from light.

【Questions】

1. After getting emerald green solution, What is the difference between cooling in an ice water bath or under ordinary room temperature?

2. What effect will hydrogen peroxide have if it is not removed by heating?

Risk Assessment and Control Worksheet

TASK	Preparation of tripotassium trioxalatoferrate				Assessor	Zeying WU, Hongsong WANG, Xingxing WU					Date	1 JUL. 2021			
Process Steps	Hazard Identification or Physical Hazard				Health Risk	Safety Measures					Checklist — Risk Assessment				Action
	Chemical Name	Class	Concn.	Amount	MSDS	Type of Injury Possible	To reduce or eliminate the risk	Technique	Glasses	Gloves	Fume Cbd.	Likely 1~4	Hazard 1~4	Risk Score	Risk Level H M L	In case of Emergency/Spill Special Comments
I) Weigh 4.0 g of $FeSO_4 \cdot 7H_2O$ crystals, put them into a beaker and add 15 mL of deionized water, heat to dissolve.	Ferrous sulfate	—	—	4.0 g	✓	Skin and eye irritant	Handle with care		✓	✓		1	1	1	L	Sponge & water
II) Add 1 mol·L^{-1} $H_2C_2O_4$ 20 mL, stir and heat to boil to form a light yellow precipitate FeC_2O_4, wash the precipitate 3 times by decantation, and use 25 mL H_2O each time to wash away soluble impurities.	$H_2C_2O_4$	—	1 M	20 mL	✓	Swallowing/ skin contact harmful	—		✓	✓		1	2	2	L	Sponge & water
III) 10 mL of saturated $K_2C_2O_4$ solution was added to the above precipitate, Heated to 40 ℃ in a water bath.	Potassium oxalate	—	saturated solution	10 mL	✓	Swallowing harmful	Handle with care		✓	✓		1	2	2	L	Report immediately to Demonstrator or Prep Room
IV) 20 mL of a 3% H_2O_2 solution was added dropwise. The solution was continuously stirred and maintained at a temperature of about 40 ℃ to fully oxidize Fe(II) to Fe(III).	H_2O_2	5.1+8	3%	20 mL	✓	Skin burns and eye injuries	Handle with care		✓	✓		1	3	3	L	Sponge & water
V) After the addition was complete, the solution was heated to boiling to remove excess H_2O_2.	heating device	—				Electric shock & scald	Ensure cord is tagged	✓	✓			1	3	3	L	Report immediately to Demonstrator or Prep Room

continued

Risk Assessment and Control Worksheet

TASK	Preparation of tripotassium trioxalatoferrate					Assessor	Zeying WU, Hongsong WANG, Xingxing WU				Date	1 JUL. 2021		
Process Steps	Hazard Identification					Health Risk		Safety Measures				Checklist~Risk Assessment		Action
	Chemical Name	Class or Physical Hazard	Concn.	Amount	MSDS	Type of Injury Possible	To reduce or eliminate the risk	Technique	Glasses	Gloves	Fume Cbd.	Likely	Hazard	Risk Score	Risk Level	In case of Emergency/Spill	Special Comments
VI) To maintain the near-boiling state of the precipitate, first add 1 mol·L^{-1} H$_2$C$_2$O$_4$ 7 mL, and then dropwise add 1 mol·L^{-1} H$_2$C$_2$O$_4$ 1~2 mL while hot to dissolve the precipitate. The pH value of the green solution is maintained at 4~5, filter the solution into a 150 mL beaker while it is still hot, and control the filtrate to about 30 mL. Cool to room temperature, crystallize, and filter to dryness to obtain potassium iron (Ⅲ) trioxalate crystals. Weigh, calculate the yield, and store the crystals in a desiccator protected from light.	H$_2$C$_2$O$_4$	—	1 M	20 mL		Swallowing/ skin contact harmful	Handle with care		√	√		1	2	2	L	Sponge & water	
												1~4	1~4		H M L		
Experimental Product (s)	Potassium trioxalatoferrate	—		2.0 g	√	—	Handle with care		√	√		1	1	1	L	Sponge & water	
Experimental Waste (s)																	
Final assessment of overall process.	Low risk when correct procedures followed																
Signatures	Supervisor:					Lab Manager/OHS Representative:						Student/Operator:					

实验 23　铬(Ⅲ)配合物的制备和分裂能的测定

【实验目的】

1. 掌握 Cr 配合物的合成方法。
2. 熟悉 Cr 的价态变化。
3. 掌握分光光度计的使用。
4. 熟悉晶体场理论。

【实验原理】

根据配位化学理论,不同类型配体靠近配位中性离子和原子时,由于配体给电子能力不同,使得中心离子和原子的 d 轨道分裂形成了两组不同能量的轨道 e_g 和 t_{2g} 轨道。根据光谱测试,可计算两组轨道之间的能差 Δ。

【器材与试剂】

1. 器材:烧杯,玻璃棒,天平,滴管,水浴锅,蒸发皿,酒精灯,三脚架等。
2. 试剂:草酸钾,重铬酸钾,草酸,乙二胺四乙酸钠,氯化铬,硫酸铬钾,丙酮。

【实验内容】

称取 0.5 g 重铬酸钾,用研钵碾细后溶于 10 mL 去离子水中,加热使其溶解。称取 1.2 g 草酸和 0.6 g 草酸钾,加热搅拌溶解。得到棕色溶液。将其转移到蒸发皿中,加热蒸发,颜色不断加深。等溶液出现晶膜后,停止加热,冷却。注意,溶液不可过分蒸干,否则目标产品将粘在蒸发皿中。过滤,得到墨绿色固体,用丙酮洗涤数次后,烘干,即得目标产品 $K_3[Cr(C_2O_4)_3]$ 晶体,称量计算产率。

称取 0.1 g $K_3[Cr(C_2O_4)_3]$ 晶体,溶于 20 mL 去离子水中,配制成溶液 a。

称取 0.4 g 硫酸铬钾,溶于 20 mL 去离子水中,配成溶液 b。

称取 0.1 g 三氯化铬和 0.14 g EDTA 溶于 50 mL 去离子水中,配成溶液 c。

将配制好的 a,b,c 三种 Cr 配合物溶液进行吸光度测试,设定分光光度计测试范围为 350~700 nm,以去离子水为基准溶液,并每隔 10 nm 测定一次吸光度,并记录数据(如下表所示)。

表 23-1　实验数据记录表

波长/nm	$[Cr(C_2O_4)_3]^{3-}$	$[Cr(H_2O)_6]^{3+}$	$Cr(EDTA)^-$
350			
360			

续 表

波长/nm	[Cr(C₂O₄)₃]³⁻	[Cr(H₂O)₆]³⁺	Cr(EDTA)⁻
370			
380			
…			

根据表格所得数据,找出最大吸收波长 λ_{max},并计算各配合物的晶体场分裂能 Δ(计算公式如下,Δ 单位:cm^{-1})。

$$\Delta = \frac{1}{\lambda_{max}} \times 10^7$$

【注意事项】

该实验中使用重铬酸钾等物质对人体伤害较大,建议在通风性能良好的实验场所进行,必要时请佩戴防护用具。

【思考题】

1. $K_3[Cr(C_2O_4)_3]$ 在溶液中的浓度对 λ_{max} 造成什么影响?
2. 三种溶液 a,b,c 的 λ_{max} 分别多少?有何规律?

● 视频演示

风险评估及控制工作表

铬(Ⅲ)配合物的制备和分裂能的测定

评估人：吴泽颖、王红松、吴星星　　评估日期：2021年7月1日

实验任务															
过程步骤	化合物名称	危害鉴别 类别	危害鉴别 浓度	危害鉴别 用量	化学品安全技术说明书	健康风险 可能的伤害种类	降低或消除风险	安全措施 技术	安全措施 防护眼镜	安全措施 手套	安全措施 通风橱	清单/风险评估	风险评估 可能性 1~4	风险评估 危害 1~4	风险评估 风险	危害级别 高中低	紧急情况/泼洒特殊备注
Ⅰ) 称取0.5 g 重铬酸钾，用研钵碾细后溶于10 mL 去离子水中，加热使其溶解。称取1.2 g 草酸和0.6 g 草酸钾，加热搅拌溶解。	重铬酸钾	9	—	0.5 g	√	吞咽会中毒皮肤灼伤和眼损伤	关注		√	√			1	2	3	低	立即向实验老师或准备实验人员报告
	草酸	—	—	1.2 g	√	吞咽/皮肤接触有害	关注		√	√			1	2	2	低	立即向实验老师或准备实验人员报告
	草酸钾	—	—	0.6 g	√	吞咽有害	关注		√	√			1	2	2	低	立即向实验老师或准备实验人员报告
Ⅱ) 得到棕色溶液。将其转移到蒸发皿中，加热蒸发，等溶液出现晶膜后，停止加热，冷却。注意，溶液不可过分蒸干，否则目标产品将粘在蒸发皿中，得到墨绿色固体，用丙酮洗涤数次后，烘干，即得目标产品 $K_3[Cr(C_2O_4)_3]$ 晶体，称量计算产率。	加热装置					电击和烫伤	确保电线贴上标签	√	√				1	3	3	低	立即向实验老师或准备实验人员报告
	丙酮	3	—	适量	√	严重眼刺激	关注		√	√			1	4	4	低	大量水冲洗
Ⅲ) 称取0.1 g $K_3[Cr(C_2O_4)_3]$ 晶体，溶于20 mL 去离子水中，配制成溶液a。	草酸铬钾	—	—	0.1 g	√	皮肤/眼刺激	关注		√	√			1	2	2	低	大量水冲洗

续表

风险评估及控制工作表

实验任务：铬(Ⅲ)配合物的制备和分裂能的测定　　评估人：吴泽颖，王红松，吴星星　　评估日期：2021年7月1日

过程步骤	化合物名称	危害鉴别				健康风险		安全措施			清单/风险评估	风险评估			紧急情况/泼洒措施	特殊备注
		类别	浓度	用量	化学品安全技术说明书	可能的伤害种类	降低或消除风险	技术	防护眼镜	手套	通风橱	可能性	危害	风险	危害级别	
		或物理危害										1~4	1~4		高中低	
Ⅳ) 称取 0.4 g 硫酸铬钾，溶于 20 mL去离子水中，配成溶液 b。	硫酸铬钾	—	—	0.4 g	√	皮肤/眼刺激	关注		√	√		1	2	2	低	大量水冲洗
Ⅴ) 称取 0.1 g 三氯化铬和0.14 g EDTA 溶于 50 mL 去离子水中，配成溶液 c。	三氯化铬	—	—	0.1 g	√	皮肤过敏	关注		√	√		1	1	1	低	大量水冲洗
	EDTA	—	—	0.14 g	√	皮肤/眼刺激	关注		√	√		1	1	1	低	大量水冲洗
实验产物	草酸铬钾	—	—	1.3 g	√	皮肤刺激	关注		√	√		1	1	1	L	用大量水冲洗
实验废弃物																
整个过程的最终评估	低风险（按照正确流程操作）															

签名　　实验老师：　　实验室主任/职业安全健康(OHS)代表：　　学生/操作人：

Experiment 23　Preparation of chromium (Ⅲ) complex and determination of splitting energy

【Objectives】

1. Master synthesis method of Cr complexes.
2. Familiar with the valence change of Cr.
3. Master the operation of spectrophotometer.
4. Familiar with crystal field theory.

【Principles】

According to the theory of coordination chemistry, when different types of ligands are closed to coordinating neutral ions or atoms, due to the different electron donating abilities of the ligands, the d orbital split of the central ions or the atoms splite into two groups of orbits: e_g and t_{2g} with different energies. Based on the spectral test, the splitting energy Δ between the two sets of orbitals can be calculated.

【Apparatus and Chemicals】

1. Apparatus

Beaker; glass rod; balance; dropper; water bath; evaporator; alcohol lamp; tripod; etc.

2. Chemicals

Potassium oxalate; potassium dichromate; oxalic acid; sodium ethylenediamine tetraacetate; chromium chloride; potassium potassium sulfate; acetone.

【Experimental Procedures】

Weigh 0.5 g of potassium dichromate and grind it, heat to dissolve it in 10 mL deionized water. Weigh 1.2 g of oxalic acid and 0.6 g of potassium oxalate, heat and dissolve in above solution. A brown solution was obtained. Transfer it to an evaporating dish, heat and evaporate it, and the color will become deep. After the solution appears a crystal film, stop heating and cool. Be careful not to over-evaporate the solution, because the target product will stick to the evaporating dish. Filtration to obtain a dark green solid, washed several times with acetone, and dried to obtain the target product $K_3[Cr(C_2O_4)_3]$ crystals, weighed to calculate the yield.

Weigh 0.1 g $K_3[Cr(C_2O_4)_3]$ crystals, dissolve them in 20 mL deionized water, and prepare solution a.

Weigh 0.4 g of potassium chromium sulfate and dissolve it in 20 mL deionized water to make solution b.

Weigh 0.1 g of chromium trichloride and 0.14 g of EDTA into 50 mL of deionized water to prepare solution c.

a, b, and c, three Cr complex solutions were prepared for absorbance test. The spectrophotometer test range was set to 350~700 nm. Deionized water was used as the reference solution. The absorbance was measured every 10 nm and the data was recorded (As shown in the table below).

Table 23 - 1 Experimental data

Wavelength /nm	$[Cr(C_2O_4)_3]^{3-}$	$[Cr(H_2O)_6]^{3+}$	$Cr(EDTA)^-$
350			
360			
370			
380			
...			

According to the data obtained from the table, find the maximum absorption wavelength λ_{max} and calculate the crystal field splitting energy Δ of each complex (the calculation formula is as follows, Δ unit: cm^{-1}).

$$\Delta = \frac{1}{\lambda_{max}} \times 10^7$$

【Notes】

The potassium dichromate is harmful to the human body. It is recommended to carry out in a well-ventilated experimental place. Wear protective equipment if necessary.

【Questions】

1. What effect does the concentration of $K_3[Cr(C_2O_4)_3]$ in solution have on λ_{max} value?
2. What are the λ_{max} of the three solutions a, b, and c, and what are the rules?

Risk Assessment and Control Worksheet

TASK	Preparation of chromium (III) complex and determination of splitting energy					Assessor	Zeying WU, Hongsong WANG, Xingxing WU						Date	1 JUL. 2021			
Process Steps	Hazard Identification					Health Risk		Safety Measures				Checklist — Risk Assessment	Risk Assessment			Action	
	Chemical Name	Class or Physical Hazard	Concn.	Amount	MSDS	Type of Injury Possible	To reduce or eliminate the risk	Technique	Glasses	Gloves	Fume Cbd.		Likely 1~4	Hazard 1~4	Risk Score	Risk Level H M L	In case of Emergency/Spill / Special Comments
I) Weigh 0.5 g of potassium dichromate and grind it, heat to dissolve it in 10 mL deionized water. Weigh 1.2 g of oxalic acid and 0.6 g of potassium oxalate, heat and dissolve in above solution. A brown solution was obtained.	Potassium dichromate	9		0.5 g	√	Swallowing harmful Skin burns and eye injuries	Handle with care		√	√			1	2	3	L	Report immediately to Demonstrator or Prep Room
	Oxalic acid	—		1.2 g	√	Swallowing/ skin contact harmful	Handle with care		√	√			1	2	2	L	Report immediately to Demonstrator or Prep Room
	Potassium oxalate	—		0.6 g	√	Swallowing harmful	Handle with care		√	√			1	2	2	L	Report immediately to Demonstrator or Prep Room
II) Transfer it to an evaporating dish, heat and evaporate it, and the color will become deep. After the solution appears a crystal film, stop heating and cool. Be careful not to over-evaporate the solution, because the target product will stick to the evaporating dish. Filtration to obtain a dark green solid, washed several times with acetone, and dried to obtain the target product $K_3[Cr(C_2O_4)_3]$ crystals, weighed to calculate the yield.	Heating device					Electric shock or scald	Ensure cord is tagged	√	√				1	3	3	L	Report immediately to Demonstrator or Prep Room
	Acetone	3		Little	√	Eye irritant	Handle with care		√	√			1	4	4	L	Sponge & water

continued

Risk Assessment and Control Worksheet

TASK	Preparation of chromium (Ⅲ) complex and determination of splitting energy			Assessor	Zeying WU, Hongsong WANG, Xingxing WU					Date	1 JUL. 2021		

Process Steps	Hazard Identification				Health Risk		Safety Measures				Checklist～Risk Assessment			Action
	Chemical Name	Class or Physical Hazard	Concn.	Amount	MSDS	Type of Injury Possible	To reduce or eliminate the risk	Technique	Glasses	Gloves	Fume Cbd.	Likely	Hazard	Risk Score	Risk Level H M L	In case of Emergency/Spill Special Comments

Process Steps	Chemical Name	Class	Concn.	Amount	MSDS	Type of Injury	To reduce/eliminate risk	Tech.	Glasses	Gloves	Fume	Likely 1~4	Hazard 1~4	Risk Score	Risk Level	Action
Ⅲ) Weigh 0.1 g K₃[Cr(C₂O₄)₃] crystals, dissolve them in 20 mL deionized water.	Potassium chromium oxalate	—		0.1 g	√	Skin and eye irritant	Handle with care		√	√		1	2	2	L	Sponge & water
Ⅳ) Weigh 0.4 g of potassium chromium sulfate and dissolve it in 20 mL deionized water to make solution.	Chromium potassium sulfate	—		0.4 g	√	Skin and eye irritant	Handle with care		√	√		1	2	2	L	Sponge & water
Ⅴ) Weigh 0.1 g of chromium trichloride and 0.14 g of EDTA into 50 mL of deionized water to prepare solution.	Chromium trichloride	—		0.1 g	√	Skin allergy	Handle with care		√	√		1	1	1	L	Sponge & water
	EDTA	—		0.14 g	√	Skin and eye irritant	Handle with care		√	√		1	1	1	L	Sponge & water
Experimental Product (s)	Potassium chromium oxalate	—		1.3 g	√	Skin irritant	Handle with care		√	√		1	1	1	L	Sponge & water
Experimental Waste (s)																
Final assessment of overall process	Low risk when correct procedures followed															

Signatures Supervisor: Lab Manager/OHS Representative: Student/Operator:

实验 24　氯化一氯·五氨合钴(Ⅲ)水合反应活化能的测定

【实验目的】

1. 掌握 Co 配合物的合成方法。
2. 熟悉 Co 的价态变化。

【实验原理】

根据配位化学理论，NH_3 和 Cl^- 在溶液中与钴离子配位形成配位化合物。根据配体浓度、比例不同可形成多种 Co 的配合物。

【器材与试剂】

1. 器材：烧杯，玻璃棒，天平，滴管，水浴锅，蒸发皿，酒精灯，三脚架等。
2. 试剂：氯化钴，浓氨水，浓盐酸，无水乙醇，无水丙酮，硝酸，氯化铵，过氧化氢。

【实验内容】

量取 3 mL 浓氨水，加入 0.5 g 氯化铵固体配成溶液。称取 1.0 g 氯化钴晶体，碾碎后加入上述溶液中。得到黄红色沉淀，后缓慢加入 3% 的 H_2O_2 溶液 1 mL。溶液变为深红色。再加入浓盐酸 3 mL 得到紫红色 $[CoCl(NH_3)_5]Cl_2$ 晶体。抽滤后使用乙醇和丙酮洗涤数次，烘干后即得目标化合物。

配制 Co 配合物溶液，浓度为 12 mmol·L^{-1}，其中 HNO_3 浓度为 0.3 mol·L^{-1}。溶液分成两份，分别在 60 ℃ 和 80 ℃ 水浴中加热，并每隔 5 min 测定一次吸光度，波长设定值为 550 nm，绘制曲线并计算活化能。

【注意事项】

该实验中使用浓氨水，浓盐酸和浓硝酸等物质对人体伤害较大，建议在通风性能良好的实验场所进行，必要时请佩戴防护用具。

【思考题】

1. 浓盐酸加多时，会对产品造成什么影响？
2. 若温度高于 80 ℃，试推测会有什么现象发生。

风险评估及控制工作表

实验任务	氯化一氯·五氨合钴(Ⅲ)水合反应活化能的测定				评估人	吴泽颖,王红松,吴星星			评估日期	2021年7月1日	

过程步骤	危害鉴别					健康风险		安全措施			风险评估			紧急情况/液洒特殊备注
	化合物名称	类别	浓度	用量	化学品安全技术说明书	可能的伤害种类	降低或消除风险	技术	防护眼镜	手套	通风橱	清单/风险评估	可能性 1~4	危害 1~4	风险	危害级别 高中低	措施
Ⅰ)量取3 mL 浓氨水,加入0.5 g氯化铵固体配成溶液。	浓氨水	8	17%	3 mL	√	皮肤灼伤和眼损伤	关注		√	√	√		1	3	3	低	大量水冲洗
	氯化铵	—		0.5 g	√	眼刺激	关注		√	√	√		1	1	1	低	大量水冲洗
Ⅱ)称取1.0 g氯化钴晶体,碾碎后加入上述溶液中。得到黄红色沉淀,后缓慢加入3%的H_2O_2溶液1 mL。	氯化钴	—		1.0 g	√	皮肤过敏	关注		√	√	√		1	1	1	低	大量水冲洗
	过氧化氢	5.1+8	3%	1 mL	√	皮肤灼伤和眼损伤	关注		√	√	√		1	3	3	低	大量水冲洗
Ⅲ)抽滤后使用乙醇和丙酮洗涤数次,烘干后即得目标化合物。	乙醇	3	>95%	10 mL	√	—	关注		√	√	√		1	2	2	低	大量水冲洗
	丙酮	3	>99%	10 mL	√	眼刺激	关注		√	√	√		1	3	3	低	大量水冲洗
Ⅳ)配制Co配合物溶液,浓度为12 mmol·L^{-1},其中HNO_3浓度为0.3 mol·L^{-1},溶液分成两份,分别在60℃和80℃水浴加热,并每隔5 min测定一次吸光度,波长设定值为550 nm,绘制曲线并计算活化能。	硝酸	8+5.1	0.3 M	10 mL	√	皮肤灼伤和眼损伤	关注		√	√	√		1	3	3	低	大量水冲洗
	加热装置					烫伤										低	立即报告实验老师或准备实验人员
实验产物	氯化一氯·五氨合钴(Ⅲ)	—		2.32 g	√	皮肤刺激	关注		√	√			1	1	1	L	用大量水冲洗
实验废弃物																	
整个过程的最终评估	低风险(按照正确流程操作)																

签名 实验老师: 实验室主任/职业安全健康(OHS)代表: 学生/操作人:

Experiment 24　Preparation of monochloro pentammine cobalt (Ⅲ) hydration and determination of its activation energy

【Objectives】

1. Master synthesis process of Co complexes.
2. Familiar the valence changes of Co.

【Principles】

According to the theory of coordination chemistry, NH_3 and Cl^- coordinate with Co ions in the solution to form a coordination compound. Various Co complexes can be formed depending on the ligand concentration and ratio.

【Apparatus and Chemicals】

1. Apparatus

Beaker; glass rod; balance; dropper; water bath; evaporation dish; alcohol lamp; tripod; etc.

2. Chemicals

Cobalt chloride; concentrated ammonia; concentrated hydrochloric acid; anhydrous ethanol; anhydrous acetone; nitric acid; ammonium chloride; hydrogen peroxide.

【Experimental Procedures】

Measure 3 mL of concentrated ammonia water and add 0.5 g of ammonium chloride solid to prepare a solution. 1.0 g of cobalt chloride crystals were weighed, crushed and added to the above solution. A yellow-red precipitate was obtained, and 1 mL of a 3% H_2O_2 solution was slowly added. The solution turned dark red. 3 mL of concentrated hydrochloric acid was added to obtain purple-red $[CoCl(NH_3)_5]Cl_2$ crystals. After suction filtration, it was washed several times with ethanol and acetone, and the target compound was obtained after drying.

Co complex solution was prepared at a concentration of 12 mmol $\cdot$ L^{-1}, of which HNO_3 concentration was 0.3 mol $\cdot$ L^{-1}. The solution was divided into two parts, heated in a 60 ℃ and 80 ℃ water bath, and the absorbance was measured every 5 min. The wavelength set value was 550 nm. The curve was drawn and the activation energy was calculated.

【Notes】

In this experiment, Concentrated ammonia, concentrated hydrochloric acid and concentrated nitric acid are all harmful to the human body. It is recommended to carry out in a well-ventilated experimental place. Wear protective equipment if necessary.

【Questions】

1. What effect does over adding concentrated hydrochloric acid have on the product?
2. If the temperature is higher than 80 ℃, try to predict what will happen.

Risk Assessment and Control Worksheet

TASK	Assessor	Date
Preparation of monochloro pentammine cobalt (Ⅲ) hydration and Determination of its activation energy	Zeying WU, Hongsong WANG, Xingxing WU	1 JUL. 2021

Process Steps	Hazard Identification				Health Risk	Safety Measures				Checklist↘Risk Assessment			Action	
	Chemical Name or Physical Hazard	Class	Concn.	Amount	MSDS	Type of Injury Possible	To reduce or eliminate the risk	Technique	Glasses	Gloves	Fume Cbd.	Likely 1~4	Hazard 1~4	Risk Score	Risk Level H M L	In case of Emergency/Spill Special Comments

Process Steps	Chemical Name	Class	Concn.	Amount	MSDS	Type of Injury Possible	To reduce or eliminate the risk	Technique	Glasses	Gloves	Fume Cbd.	Likely	Hazard	Risk Score	Risk Level	Action
Ⅰ) Measure 3 mL of concentrated ammonia water and add 0.5 g of ammonium chloride solid to prepare a solution.	Ammonia	8	17%	3 mL	✓	Skin burns and eye injuries	Handle with care		✓	✓	✓	1	4	4	L	Sponge & water
	Ammonium chloride	—		0.5 g	✓	Eye irritant	Handle with care		✓	✓	✓	1	1	1	L	Sponge & water
Ⅱ) 1.0 g of cobalt chloride crystals were weighed, crushed and added to the above solution. A yellow-red precipitate was obtained, and 1 mL of a 3% H_2O_2 solution was slowly added.	Cobalt chloride	—		1.0 g	✓	Skin allergy	Handle with care		✓	✓	✓	1	1	1	L	Sponge & water
	H_2O_2	5.1+8	3%	1 mL	✓	Skin burns and eye injuries	Handle with care		✓	✓	✓	1	3	3	L	Sponge & water
Ⅲ) After suction filtration, it was washed several times with ethanol and acetone, and the target compound was obtained after drying.	Ethanol	3	>95%	10 mL	✓	—	Handle with care		✓	✓	✓	1	2	2	L	Sponge & water
	Acetone	3	>99%	10 mL	✓	Eye irritant	Handle with care		✓	✓	✓	1	3	3	L	Sponge & water
Ⅳ) Co complex solution was prepared at a concentration of 12 mmol·L^{-1}, of which HNO_3 concentration was 0.3 mol·L^{-1}. The solution was divided into two parts, heated in a 60 ℃ and 80 ℃ water bath, and the absorbance was measured every 5 min.	HNO_3	8+5.1	0.3 M	10 mL	✓	Skin burns and eye injuries	Handle with care		✓	✓	✓	1	3	3	L	Sponge & water
	Heating device					Burns	Handle with care					1	2	2	低	Report immediately to Demonstrator or Prep Room

continued

Risk Assessment and Control Worksheet

TASK	Preparation of monochloro pentammine cobalt (Ⅲ) hydration and Determination of its activation energy											Assessor	Zeying WU, Hongsong WANG, Xingxing WU			Date	1 JUL. 2021	
Process Steps	Hazard Identification				Health Risk		Checklist＼Risk Assessment									Risk Assessment		Action
	Chemical Name	Class	Concn.	Amount	MSDS	Type of Injury Possible	or Physical Hazard	To reduce or eliminate the risk	Safety Measures							Likely	Hazard	Risk Score	Risk Level	In case of Emergency/Spill
									Technique	Glasses	Gloves	Fume Cbd.							Special Comments
Experimental Product (s)	Chlorine chloride・pentammine cobalt (Ⅲ)	—		2.32 g	√	Skin irritant		Handle with care		√	√				1	1～4	1	L	Sponge & water
Experimental Waste (s)																			
Final assessment of overall process	Low risk when correct procedures followed																		
Signatures	Supervisor:							Lab Manager/OHS Representative:											Student/Operator:

实验 25　多金属氧酸盐的制备及其光催化降解性能测试研究

【实验目的】

1. 了解杂多酸的结构和性质。
2. 了解钼酸盐的基本性状。
3. 掌握有机污染物的降解操作步骤。

【实验原理】

根据杂多酸盐的合成机理,Mo 系金属盐能与其他正八面体型结构形成多种拓扑结构,尤其是 Anderson 型结构中,Mo 与高碘酸根形成了八元环型结构。

【器材与试剂】

1. 器材:烧杯,玻璃棒,天平,滴管,水浴锅,蒸发皿,酒精灯,三脚架等。
2. 试剂:钼酸钠,高碘酸钠,罗丹明 B,过氧化氢。

【实验内容】

称取钼酸钠 6 g 和高碘酸钠 1 g,溶解于 30 mL 去离子水中。加热搅拌,得到白色糊状物质,后放入不锈钢水热反应釜中,180 ℃高温 1 h。取出反应釜冷却至室温,后过滤得到片状 Anderson 型 Mo 系多金属氧酸盐。

配制罗丹明 B 溶液,浓度为 40 mg·L^{-1}。称取一定量的多金属氧酸盐进行光催化降解实验,并每隔 5 min 测定一次吸光度,波长设定值为 550 nm,绘制曲线并计算降解速率。

【思考题】

1. 若在合成过程中钼酸钠比例过高会有什么影响?
2. 不加 H_2O_2,反应速率有何变化?

● 光催化降解

风险评估及控制工作表

实验任务	多金属氧酸盐的制备及其光催化降解性能测试研究				评估人	吴泽颖、王红松、吴星星				评估日期	2021年7月1日				
过程步骤	危害鉴别				健康风险		安全措施			清单/风险评估	风险评估			措施	
	化合物名称	类别	浓度	用量	化学品安全技术说明书	可能的伤害种类	降低或消除风险	技术	防护眼镜	手套	通风橱		可能性 1~4	危害 1~4	风险	危害级别 高中低	紧急情况/泼洒	特殊备注
		戒物理危害															
Ⅰ) 称取钼酸钠 6 g 和高碘酸钠 1 g,溶解于 30 mL 去离子水中。加热搅拌,得到白色糊状物质,后放入不锈钢水热反应釜中,180 ℃高温 1 h。	钼酸钠	—	—	6.0 g	√	—	关注		√	√			1	1	1	低	大量水冲洗	
	高碘酸钠	5.1	5.1	1.0 g	√	—	关注		√	√			1	1	1	低	大量水冲洗	
	加热装置	—	—	—		电击或烧伤	确保电线贴上标签	√					1	3	3	低	立即向实验老师或准备实验人员报告	
Ⅱ) 配制罗丹明 B 溶液,浓度为 40 mg·L^{-1}。称取一定量的多金属氧酸盐进行光催化降解实验。	罗丹明 B	—	40 mg·L^{-1}	40 mL	√	眼损伤	关注		√	√			1	1	1	低	大量水冲洗	
	碘酸钼多酸盐	5.1	5.1	5.6 g	√	皮肤灼伤	关注		√	√			1	2	2	低	大量水冲洗	
实验产物																		
实验废弃物																		
整个过程的最终评估	低风险(按照正确流程操作)																	

签名　实验老师:　　　　实验室主任/职业安全健康(OHS)代表:　　　　学生/操作人:

Experiment 25 Preparation of polyoxometalates and assessment of its photocatalytic degradation performance

【Objectives】

1. Understand the structure and properties of polyoxometalates.
2. Understand the properties of Mo salt.
3. Master the operation steps of degradation of organic pollutants.

【Principles】

According to the synthetic mechanism of polyoxometalates, Mo-based metal salts can form a variety of topological structures with other octahedral structures. Especially in the Anderson structure, Mo and periodic acid ions form an eight-membered ring structure.

【Apparatus and Chemicals】

1. Apparatus

Beaker, glass rod, balance, dropper, water bath, evaporation dish, alcohol lamp, tripod, etc.

2. Chemicals

Sodium molybdate, sodium periodate, Rhodamine B, hydrogen peroxide.

【Experimental Procedures】

Weigh 6 g of sodium molybdate and 1 g of sodium periodate, and dissolve in 30 mL of deionized water. Heat and stir to obtain a white paste-like substance, then put it into a stainless steel hydrothermal reaction autoclave at 180 ℃ for 1h. The autoclave was taken out and cooled to room temperature, and then filtered to obtain a flake-shaped Anderson-type polyoxometalate.

Rhodamine B solution was prepared at a concentration of 40 mg·L^{-1}. A certain amount of polyoxometalate was weighed for a photocatalytic degradation experiment, and the absorbance was measured every 5 minutes with a wavelength set value of 550 nm. A curve was drawn and the degradation rate was calculated.

【Questions】

1. If the proportion of sodium molybdate is too high in the synthesis process, what effect will it have?
2. What is the change of reaction rate without H_2O_2?

Risk Assessment and Control Worksheet

TASK	Preparation of polyoxometalates and assessment of its photocatalytic degradation performance					Assessor	Zeying WU, Hongsong WANG, Xingxing WU				Date	1 JUL. 2021		
Process Steps	Hazard Identification					Health Risk		Safety Measures			Checklist~Risk Assessment			Action
	Chemical Name	Class	Concn.	Amount	MSDS	Type of Injury Possible	To reduce or eliminate the risk	Technique	Glasses	Gloves	Fume Cbd.	Likely 1~4	Hazard 1~4	Risk Score	Risk Level H M L	In case of Emergency/Spill	Special Comments
	or Physical Hazard															
Ⅰ) Weigh 6 g of sodium molybdate and 1 g of sodium periodate, and dissolve in 30 mL of deionized water.	Sodium molybdate	—	—	6.0 g	√	—	Handle with care		√	√		1	1	1	L	Sponge & water
	Sodium periodate	5.1	—	1.0 g	√	—	Handle with care		√	√		1	1	1	L	Sponge & water
Ⅱ) Heat and stir to obtain a white paste-like substance, then put it into a stainless steel hydrothermal reaction autoclave at 180 ℃ for 1 h.	Heating device	—	—	—		Electric shock or scald	Ensure cord is tagged	√	√			1	3	3	L	Report immediately to Demonstrator or Prep Room
Ⅲ) Rhodamine B solution was prepared at a concentration of 40 mg·L⁻¹.	Rhodamine B	—	40 mg·L⁻¹	40 mL	√	Eye injuries	Handle with care		√	√		1	1	1	L	Sponge & water
Experimental Product (s)	Molybdate iodate	5.1	—	5.6 g	√	Skin burns	Handle with care		√	√		1	2	2	L	Sponge & water
Experimental Waste (s)																
Final assessment of overall process	Low risk when correct procedures followed															
Signatures	Supervisor:					Lab Manager/OHS Representative:									Student/Operator:	

269

附　录

附录1　元素的相对原子质量(A_r)表(2011)
Appendix 1　Table of standard atomic weights (2011)

元素	英文名	符号	A_r	元素	英文名	符号	A_r
银	Silver	Ag	107.868 2(2)	铒	Erbium	Er	167.259(3)
铝	Aluminum	Al	26.981 538 6(8)	铕	Europium	Eu	151.964(1)
氩	Argon	Ar	39.948(1)	氟	Fluorine	F	18.998 403 2(5)
砷	Arsenic	As	74.921 60(2)	铁	Iron	Fe	55.845(2)
金	Gold	Au	196.966 569(4)	镓	Gallium	Ga	69.723(1)
硼	Boron	B	[10.806, 10.821]	钆	Gadolinium	Gd	157.25(3)
钡	Barium	Ba	137.327(7)	锗	Germanium	Ge	72.630(8)
铍	Beryllium	Be	9.012 182(3)	氢	Hydrogen	H	[1.007 84, 1.008 11]
铋	Bismuth	Bi	208.980 40(1)	氦	Helium	He	4.002 602(2)
溴	Bromine	Br	[79.901, 79.907]	铪	Hafnium	Hf	178.49(2)
碳	Carbon	C	[12.009 6, 12.011 6]	汞	Mercury	Hg	200.592(3)
钙	Calcium	Ca	40.078(4)	钬	Holmium	Ho	164.930 32(2)
镉	Cadmium	Cd	112.411(8)	碘	Iodine	I	126.904 47(3)
铈	Cerium	Ce	140.116(1)	铟	Indium	In	114.818(1)
氯	Chlorine	Cl	[35.446, 35.457]	铱	Iridium	Ir	192.217(3)
钴	Cobalt	Co	58.933 195(5)	钾	Potassium	K	39.098 3(1)
铬	Chromium	Cr	51.996 1(6)	氪	Krypton	Kr	83.798(2)
铯	Cesium	Cs	132.905 451 9(2)	镧	Lanthanum	La	138.905 47(7)
铜	Copper	Cu	63.546(3)	锂	Lithium	Li	[6.938, 6.997]
镝	Dysprosium	Dy	162.500(1)	镥	Lutetium	Lu	174.966 8(1)

续 表

元素	英文名	符号	A_r	元素	英文名	符号	A_r
镁	Magnesium	Mg	[24.304, 24.307]	锑	Antimony	Sb	121.760(1)
锰	Manganese	Mn	54.938 045(5)	钪	Scandium	Sc	44.955 912(6)
钼	Molybdenum	Mo	95.96(2)	硒	Selenium	Se	78.96(3)
氮	Nitrogen	N	[14.006 43, 14.007 28]	硅	Silicon	Si	[28.084, 28.086]
钠	Sodium	Na	22.989 769 28(2)	钐	Samarium	Sm	150.36(2)
铌	Niobium	Nb	92.906 38(2)	锡	Tin	Sn	118.710(7)
钕	Neodymium	Nd	144.242(3)	锶	Strontium	Sr	87.62(1)
氖	Neon	Ne	20.179 7(6)	钽	Tantalum	Ta	180.947 88(2)
镍	Nickel	Ni	58.693 4(4)	铽	Terbium	Tb	158.925 35(2)
氧	Oxygen	O	[15.999 03, 15.999 77]	碲	Tellurium	Te	127.60(3)
锇	Osmium	Os	190.23(3)	钍	Thorium	Th	232.038 06(2)
磷	Phosphorus	P	30.973 762(2)	钛	Titanium	Ti	47.867(1)
镤	Protactinium	Pa	231.035 88(2)	铊	Thallium	Tl	[204.382, 204.385]
铅	Lead	Pb	207.2(1)	铥	Thulium	Tm	168.934 21(2)
钯	Palladium	Pd	106.42(1)	铀	Uranium	U	238.028 91(3)
镨	Praseodymium	Pr	140.907 65(2)	钒	Vanadium	V	50.941 5(1)
铂	Platinum	Pt	195.084(9)	钨	Tungsten	W	183.84(1)
铷	Rubidium	Rb	85.467 8(3)	氙	Xenon	Xe	131.293(6)
铼	Rhenium	Re	186.207(1)	钇	Yttrium	Y	88.905 85(2)
铑	Rhodium	Rh	102.905 50(2)	镱	Ytterbium	Yb	173.054(5)
钌	Ruthenium	Ru	101.07(2)	锌	Zinc	Zn	65.38(2)
硫	Sulfur	S	[32.059, 32.076]	锆	Zirconium	Zr	91.224(2)

注：(1) 括号内的数字指末位数字的不确定度。
(2) 表中数据引自文献：Pure Appl Chem. Vol.85, No.5, 1047-1078. 2013.

附录 2 常用酸、碱溶液的物理性质
Appendix 2 Physical properties of common acids and bases

试剂名称	英文名称	化学式	密度 ρ(g/mL)	质量分数 ω(%)	物质的量浓度 c(mol·L^{-1})
盐酸	Hydrochloric Acid	HCl	1.18~1.19	36~38	11.7~12.4
硝酸	Nitric Acid	HNO$_3$	1.39~1.40	65.0~68.0	14.4~15.3
硫酸	Sulfuric Acid	H$_2$SO$_4$	1.83~1.84	95~98	17.8~18.4
磷酸	Phosphoric Acid	H$_3$PO$_4$	1.69	85	14.6
高氯酸	Perchloric Acid	HClO$_4$	1.68	70.0~72.0	11.7~12.0
冰醋酸	Glacial Acetia Acid	HAc	1.05	GR,99.8;AR,99.5;CP,99.0	17.4
氢氟酸	Hydrofluoric Acid	HF	1.13	40	22.5
氢溴酸	Hydrobromic Acid	HBr	1.49	47.0	8.6
氨水	Ammonia	NH$_3$·H$_2$O	0.88~0.90	25.0~28.0	13.3~14.8

附录3 常用弱电解质的解离常数
Appendix 3 Dissociation constants of common weak electrolytes

试剂	英文名	化学式	$K^{\ominus}$	$pK^{\ominus}$
醋酸	Acetic acid	HAc	1.76×10^{-5}	4.75
碳酸	Carbonic acid	H_2CO_3	$K_1=4.30\times10^{-7}$	6.37
			$K_2=5.61\times10^{-11}$	10.25
草酸	Oxalic acid	$H_2C_2O_4$	$K_1=5.90\times10^{-2}$	1.23
			$K_2=6.40\times10^{-5}$	4.19
亚硝酸	Nitrous acid	HNO_2	4.6×10^{-4} (285.5 K)	3.37
磷酸	Phosphoric acid	H_3PO_4	$K_1=7.52\times10^{-3}$	2.12
			$K_2=6.23\times10^{-8}$	7.21
			$K_3=2.2\times10^{-13}$ (291 K)	12.67
亚硫酸	Sulphuric acid	H_2SO_3	$K_1=1.54\times10^{-2}$ (291 K)	1.81
			$K_2=1.02\times10^{-7}$	6.91
硫酸	Sulphuric acid	H_2SO_4	$K_2=1.20\times10^{-2}$	1.92
氢硫酸	Hydrosulphuric acid	H_2S	$K_1=9.1\times10^{-8}$ (291 K)	7.04
			$K_2=1.1\times10^{-12}$	11.96
氢氰酸	Hydrocyanic acid	HCN	4.93×10^{-10}	9.31
铬酸	Chromic acid	H_2CrO_4	$K_1=1.8\times10^{-1}$	0.74
			$K_2=3.20\times10^{-7}$	6.49
硼酸	Boric acid	H_3BO_3	5.8×10^{-10}	9.24
氢氟酸	Hydrofluoric acid	HF	3.53×10^{-4}	3.45
过氧化氢	Hydrogen peroxide	H_2O_2	2.4×10^{-12}	11.62
次氯酸	Hypochloric acid	HClO	2.95×10^{-5} (291 K)	4.53
碘酸	Iodic acid	HIO_3	1.69×10^{-1}	0.77
砷酸	Arsenic acid	H_3AsO_4	$K_1=5.62\times10^{-3}$ (291 K)	2.25
			$K_2=1.70\times10^{-7}$	6.77
			$K_3=3.95\times10^{-12}$	11.40
亚砷酸	Arsenious acid	$HAsO_2$	6×10^{-10}	9.22
铵离子	Ammonium ion	NH_4^+	5.66×10^{-10}	9.25
氨水	Ammonia	$NH_3\cdot H_2O$	1.79×10^{-5}	4.75

续 表

试剂	英文名	化学式	$K^{\ominus}$	p$K^{\ominus}$
联氨	Hydrazine	N_2H_4	8.91×10^{-7}	6.05
羟氨	Hydroxylamine	NH_2OH	9.12×10^{-9}	8.04
氢氧化铅	Lead Hydroxide	$Pb(OH)_2$	9.6×10^{-4}	3.02
氢氧化锂	Lithium Hydroxide	LiOH	6.31×10^{-1}	0.2
氢氧化铍	Beryllium Hydroxide	$Be(OH)_2$	1.78×10^{-6}	5.75
		$BeOH^+$	2.51×10^{-9}	8.6
氢氧化铝	Aluminum hydroxide	$Al(OH)_3$	5.01×10^{-9}	8.3
		$Al(OH)_2^+$	1.99×10^{-10}	9.7
氢氧化锌	Zinc hydroxide	$Zn(OH)_2$	7.94×10^{-7}	6.1
甲酸	Formic acid	HCOOH	1.77×10^{-4}(293 K)	3.75
柠檬酸	Citric acid	$(HOOCCH_2)_2C(OH)COOH$	$K_1=7.1\times10^{-4}$	3.14
			$K_2=1.68\times10^{-5}$(293 K)	4.77
			$K_3=4.1\times10^{-7}$	6.39
酒石酸	Tartaric acid	$HOOC(CHOH)_2COOH$	$K_1=1.04\times10^{-3}$	2.98
			$K_2=4.55\times10^{-5}$	4.34
二胺四乙酸	Diamine Tetraacetic acid		$K_1=1.0\times10^{-2}$	2.0
			$K_2=2.1\times10^{-3}$	2.68
			$K_3=6.9\times10^{-7}$	6.16
			$K_4=5.9\times10^{-11}$	10.23

附录 4　常用酸碱指示剂

指示剂名称	变色 pH 范围与颜色变化	配制方法
甲基紫(第一变色范围)	0.13～0.5;黄～绿	1 g·L^{-1} 或 0.5 g·L^{-1} 水溶液
甲酚红(第一变色范围)	0.2～1.8;红～黄	0.04 g 指示剂溶于 100 mL 50%乙醇
甲基紫(第二变色范围)	1.0～1.5;绿～蓝	1 g·L^{-1} 水溶液
百里酚蓝(第一变色范围)	1.2～2.8;红～黄	1 g 指示剂溶于 100 mL 20%乙醇
甲基紫(第三变色范围)	2.0～3.0;蓝～紫	1 g·L^{-1} 水溶液
甲基橙	3.1～4.4;红～黄	1 g·L^{-1} 水溶液
溴酚蓝	3.0～4.6;黄～蓝	1 g 指示剂溶于 100 mL 20%乙醇
刚果红	3.0～5.2;蓝紫～红	1 g·L^{-1} 水溶液
溴甲酚绿	3.8～5.4;黄～蓝	0.1 g 指示剂溶于 100 mL 20%乙醇
甲基红	4.4～6.2;红～黄	0.1 g 或 0.2 g 指示剂溶于 100 mL 60%乙醇
溴酚红	5.0～6.8;黄～红	0.1 g 或 0.04 g 指示剂溶于 100 mL 20%乙醇
溴百里酚蓝	6.0～7.6;黄～蓝	0.05 g 指示剂溶于 100 mL 20%乙醇
中性红	6.8～8.0;红～亮黄	0.1 g 指示剂溶于 100 mL 60%乙醇
酚红	6.8～8.0;黄～红	0.1 g 指示剂溶于 100 mL 20%乙醇
甲酚红	7.2～8.8;亮黄～紫红	0.1 g 指示剂溶于 100 mL 50%乙醇
百里酚蓝(第二变色范围)	8.0～9.0;黄～蓝	1 g 指示剂溶于 100 mL 20%乙醇
酚酞	8.0～9.6;无色～紫红	0.1 g 指示剂溶于 100 mL 60%乙醇
百里酚酞	9.4～10.6;无色～蓝	0.1 g 指示剂溶于 100 mL 90%乙醇

Appendix 4 Common acid-base indicators

Name	pH Range of Color Change and Color Change	Preparation Method
Methyl Violet (scope of the first color change)	0.13~0.5; Yellow~Green	1 g·L^{-1} or 0.5 g·L^{-1} aqueous solution
Cresol Red (scope of the first color change)	0.2~1.8; Red~Yellow	0.04 g indicator dissolved in 100 mL 50% ethanol
Methyl Violet (scope of the second color change)	1.0~1.5; Green~Blue	1 g·L^{-1} aqueous solution
Thymol Blue (scope of the first color change)	1.2~2.8; Red~Yellow	1 g indicator dissolved in 100 mL of 20% ethanol
Methyl Violet (scope of the third color change)	2.0~3.0; Blue~Purple	1 g·L^{-1} aqueous solution
Methyl Orange	3.1~4.4; Red~Yellow	1 g·L^{-1} aqueous solution
Bromphenol Blue	3.0~4.6; Yellow~Blue	1 g indicator dissolved in 100 mL of 20% ethanol
Congo Red	3.0~5.2; Blue violet~Red	1 g·L^{-1} aqueous solution
Bromocresol Green	3.8~5.4; Yellow~Blue	0.1 g indicator dissolved in 100 mL of 20% ethanol
Methyl Red	4.4~6.2; Red~Yellow	0.1 g or 0.2 g indicator dissolved in 100 mL of 60% ethanol
Bromphenol Red	5.0~6.8; Yellow~Red	0.1 g or 0.04 g indicator dissolved in 100 mL of 20% ethanol
Bromothymol Blue	6.0~7.6; Yellow~Blue	0.05 g indicator dissolved in 100 mL of 20% ethanol
Neutral Red	6.8~8.0; Red~Bright Yellow	0.1 g indicator dissolved in 100 mL of 60% ethanol
Phenol Red	6.8~8.0; Yellow~Red	0.1 g indicator dissolved in 100 mL of 20% ethanol
Cresol Red	7.2~8.8; Bright Yellow~purple Red	0.1 g indicator dissolved in 100 mL of 50% ethanol
Thymol Blue (scope of the second color change)	8.0~9.0; Yellow~Blue	1 g indicator dissolved in 100 mL of 20% ethanol
Phenolphthalein	8.0~9.6; Colorless~Purple Red	0.1 g indicator dissolved in 100 mL of 60% ethanol
Thyme Phthalein	9.4~10.6; Colorless~Blue	0.1 g indicator dissolved in 100 mL of 90% ethanol

附录 5 常用缓冲溶液

组成	pK_a	缓冲液 pH	配制方法
$NH_2CH_2COOH-HCl$	$2.35(pK_{a1})$	2.3	取氨基乙酸 150 g 溶于 500 mL 水中,加浓 HCl 80 mL,加水稀释至 1 L
H_3PO_4-枸橼酸盐		2.5	取 $Na_2HPO_4 \cdot 12H_2O$ 113 g 溶于 200 mL 水中,加柠檬酸 387 g,溶解,过滤后,稀释至 1 L
$C_2H_3ClO_2-NaOH$	2.86	2.8	取一氯乙酸 200 g 溶于 200 mL 水中,加 NaOH 40 g,溶解后,稀释至 1 L
$C_8H_5KO_4-HCl$	$2.95(pK_{a1})$	2.9	取邻苯二甲酸氢钾 500 g 溶于 500 mL 水中,加浓 HCl 80 mL,稀释至 1 L
$HCOOH-NaOH$	3.76	3.7	取甲酸 95 g 和 NaOH 40 g 于 500 mL 水中,溶解,稀释至 1 L
$NaAc-HAc$	4.74	4.7	取无水 NaAc 83 g 溶于水中,加冰 HAc 60 mL,稀释至 1 L
$C_6H_{12}N_4-HCl$	5.15	5.4	取六亚甲基四胺 40 g 溶于 200 mL 水中,加浓 HCl 10 mL,稀释至 1 L
Tris－HCl [三羟甲基氨甲烷 $CNH_2=(HOCH_2)_3$]	8.21	8.2	取 Tris 试剂 25 g 溶于水中,加浓 HCl 8 mL,稀释至 1 L
NH_3-NH_4Cl	9.26	9.2	取 NH_4Cl 54 g 溶于水中,加浓氨水 63 mL,稀释至 1 L

Appendix 5 Common buffer solutions

Composition	pK_a	缓冲液 pH	配制方法
NH_2CH_2COOH - HCl	2.35(pK_{a1})	2.3	150 g of glycine dissolved in 500 mL of water, add 80 mL of concentrated hydrochloric acid, diluted to 1 L
H_3PO_4 - Citrate		2.5	113 g of $Na_2HPO_4 \cdot 12H_2O$ dissolved in 200 mL of water, add 387 g of citric acid, dissolved, filtered, and diluted to 1 L
$C_2H_3ClO_2$ - NaOH	2.86	2.8	200 g of chloroacetic acid dissolved in 200 mL of water, add 40 g of NaOH, dissolved, and diluted to 1 L
$C_8H_5KO_4$ - HCl	2.95(pK_{a1})	2.9	500 g of potassium acid phthalate dissolved in 500 mL of water, add 80 mL of concentrated hydrochloric acid, diluted to 1 L
HCOOH - NaOH	3.76	3.7	95 g of formic acid and 40 g of sodium hydroxide dissolved in 500 mL of water, dissolved, and diluted to 1 L
NaAc - HAc	4.74	4.7	83 g of anhydrous sodium acetate dissolved in water, add 60 mL of glacial acetic acid, diluted to 1 L
$C_6H_{12}N_4$ - HCl	5.15	5.4	40 g of hexamethylenetetramine dissolved in 200 mL of water, add 10 mL of concentrated hydrochloric acid, diluted to 1 L
Tris - HCl	8.21	8.2	25 g of Tris dissolved in water, add 8 mL of concentrated hydrochloric acid, diluted to 1 L
NH_3 - NH_4Cl	9.26	9.2	54 g of ammonium chloride dissolved in water, add 63 mL of concentrated ammonia water, diluted to 1 L

参考文献

[1] 李荣.无机与分析化学实验[M].北京:机械工业出版社,2014.
[2] 王新宏.分析化学实验(双语版)[M].北京:科学出版社,2009.
[3] 黄应平.分析化学实验(英汉双语教材)[M].武汉:华中师范大学出版社,2012.
[4] 赵滨,马林,沈建中等.无机化学与化学分析实验[M].上海:复旦大学出版社,2008.
[5] 刘冰,徐强.无机与分析化学实验.北京:化学工业出版社,2015.
[6] 华中师范大学,东北师范大学,陕西师范大学等.分析化学实验(第四版)[M].北京:高等教育出版社,2015.
[7] 邓乐平.氢氧化钠标准溶液浓度不确定度的评定[J].化学分析计量,2004,13(3):1-5.
[8] 王佳蓉.标定高锰酸钾标准溶液浓度中影响因素的探讨[J].应用研究,2005,2:72-75.
[9] 华东理工大学,成都科技大学.分析化学[M].北京:高等教育出版社,1995.
[10] 林新华.分析化学实验指导[M].厦门:厦门大学出版社,2014.
[11] 陆荣.高锰酸钾标准滴定溶液配制与标定的影响因素[J].中国石油和化工标准与质量,2011,6:32.
[12] 于旭霞,戚彤等.高锰酸钾法与硫酸铈法测定双氧水中过氧化氢含量的对比[J].甘肃石油和化工,2012,2:51-54.
[13] 乔凤霞,康永胜,王梦歌.无汞法测定铁矿石中铁含量的实验改进[J].保定学院学报,2008,21(4):11-12.
[14] 华中师范大学,东北师范大学.分析化学实验(第3版)[M].北京:高等教育出版社,2001.
[15] 赵瑞兰,马铭,谢青季.改进的$SnCl_2$-$TiCl_3$法测定铁含量[J].实验室研究与探索,2004,8:32-33.
[16] 武汉大学.分析化学实验(第3版)[M].北京:高等教育出版社,1994.
[17] 伍惠玲.对铜盐中铜含量的准确测定[J].科技经济市场,2006,41.
[18] 四川大学化工学院,浙江大学化学系.分析化学实验(第3版)[M].北京:高等教育出版社,2003,115-118.
[19] 周静.对《铜盐中铜含量测定》实验方法的探讨[J].焦作大学学报(综合版).1996,3:61-62.
[20] 肖玉萍,张旭,曹宏杰.碘量法测定铜精矿中铜[J].光谱实验室,2011,5(28):2317-2319.
[21] 中华人民共和国国家标准.铜精矿化学分析方法--碘量法测铜[S].GB/T 1-2000.北京:中国标准出版社,2001.
[22] 胡红梅.改进可溶性氯化物中氯含量测定方法的探讨[J].天中学刊,2000,15(2):96-97.
[23] 佟琦,高丽华.莫尔法与自动电位滴定法测定水中氯离子含量的比较[J].工业水处理,2008,11(28):69-71.
[24] 武汉大学.无机及分析化学实验(第2版)[M].武汉:武汉大学出版社,2001.
[25] 焦立为.莫尔法沉淀滴定测定肥料中氯化物含量[J].理化检验(化学分册).2006,42(3):219-220.
[26] 卢爱党,王铁男等.对比实验在无机及分析化学实验教学中的应用—以银量法测定氯含量实验为例[J].化学教育(中英文),2019,8(40):42-45.
[27] 南京大学.无机及分析化学实验(第5版)[M].北京:高等教育出版社,2015.
[28] 牟文生,大连理工无机化学教研室.无机化学实验(第3版)[M].北京:高等教育出版社,2012.